The SOLAR ELECTRIC HOUSE

A Design Manual for Home-Scale Photovoltaic Power Systems

by STEVEN J. STRONG
with William G. Scheller

Rodale Press, Emmaus, Pennsylvania

To my mother, the artist, and my father, the electrician,
for providing the spark long ago.

To Marilyn and Hunter
for keeping that spark alive.

To my clients, who are the true pioneers.

Copyright © 1987 by Steven J. Strong

All rights reserved. No part of this publication may be reproduced or transmitted in any form or by any means, electronic or mechanical, including photocopy, recording, or any information storage and retrieval system, without the written permission of the publisher.

Printed in the United States of America on recycled paper containing a high percentage of de-inked fiber.

Senior Editor: Ray Wolf
Editor: Larry McClung
Copy Editor: Jane Sherman

Library of Congress Cataloging-in-Publication Data

Strong, Steven J.
The solar electric house.

Bibliography: p.
Includes index.
1. Solar houses. 2. Photovoltaic power generation.
I. Scheller, William. II. Title.
TH7414.S77 1987 697′.78 86-22061
ISBN 0–87857–647–9 hardcover

2 4 6 8 10 9 7 5 3 1 hardcover

Due to the variability of local conditions, materials, skills, site, and so forth, Rodale Press, Inc., and the Author assume no responsibility for any personal injury, property damage, or other loss of any sort suffered from the actions taken based on or inspired by information or advice presented in this book. Compliance with local codes and legal requirements is the responsibility of the User.

Contents

Foreword

Throughout the past decade, it has been impossible for anyone concerned with the quality of life on this planet to avoid speculation on the relationship between economic development and the depletion of our natural resources. It took several "energy crises" to focus our attention on this great planetary issue. Our dependence on nonrenewable sources of energy, which fueled our remarkable record of technological achievement, and the inevitable depletion of those very supplies could dramatically alter our quality of life. The temporary distraction of a politically driven "oil glut" in the late 1980s should not diminish our awareness of the problem. The rate at which fossil fuels are consumed may change, but the direction our civilization is headed will not. As surely as the sun rises, the world's oil resources will be depleted. If it isn't our own well-being that is at stake, then surely it is our children's.

Anyone building or purchasing a new home should give serious consideration to any energy device that can reduce reliance upon fossil fuel. When my wife, Wiley, and I decided to build a new house, we began to evaluate the range of energy-saving alternatives. We discussed the traditional options: extra insulation, double-glazed windows, and a "tight" house. Early in the process, however, I met Steve Strong, who extolled the virtues of photovoltaics as a means of *producing* some of the electricity to be used in my home—on-site, directly from the sun, without consumption of nonrenewable fuels.

Steve, a pioneer in energy-independent home design, explained to me how photovoltaic devices create electricity directly from sunlight. Steve does not think of photovoltaic electricity as a year-2000 possibility but as a proven and reliable technology already at work in thousands of diverse applications, providing comfort and convenience without environmental degradation and energy resource depletion. He clearly thinks these familiar twentieth-century problems *are* avoidable, and the solar electric homes he has built stand as eloquent testimony to his argument. While our photovoltaic system, which Steve designed, provides for only a portion of our electrical consumption, I am pleased to be a part of the pioneer stage of residential PV use.

At least once in every generation, our society experiences a technological breakthrough of such magnitude that we are soon unable to imagine life without it. In the 1800s it was the steam engine, the telegraph and telephone, and the introduction of electricity itself. In the last half of our century, television and computers have changed the world immeasurably. Although clearly not economically competitive with traditional sources of electrical generation at the present time, photovoltaics holds out the promise of making a similar dramatic impact upon the way we live. Requiring only sunlight as "fuel," photovoltaic devices produce electricity silently, with no depletion of materials or resources, no toxic by-products or waste of any kind, and little maintenance. Photovoltaics could become an elegant, environmentally benign solution to the problem of electrical generation—a problem heretofore exacerbated by the intractable environmental drawbacks of coal, oil, and nuclear energy.

Solar electricity enthusiasts such as Steve Strong are convinced it is the first viable, high-tech energy solution to have true universal appeal. Once it becomes cost-effective, it will become attractive to the developing as well as the developed world; it will be well suited for large central utility power plants yet also perfect for individual homes. Hopefully, our government will continue to offer tax incentives to motivate photovoltaic research and development in this country.

However, the success of the photovoltaic revolution will depend less on policies adopted by governments and utilities than on an understanding and acceptance of the value of this technology by builders, homeowners, and indeed all energy consumers. Unfortunately, the building industry often shies away from new technology. For this reason I am sometimes amazed that we have indoor plumbing or electricity in our homes at all.

Given the conservative nature of the construction industry, this book by an architect and engineer represents a major step in bringing the photovoltaic revolution closer to reality. Use the book to learn how solar electricity will figure into your future energy decisions as a consumer. Read it, as well, as a valuable case history illustrating American ingenuity at work in search of a better tomorrow.

Gilbert M. Grosvenor
President, National Geographic Society

Acknowledgments

In the course of researching and writing this book, hundreds of individuals, corporations, and research organizations were contacted, and many made significant contributions of time, photographs, and information. While it is not possible to list them all, there are a number of individuals without whose interest and support this book would have not been written.

I would like to thank the professionals who generously invested their time to provide comments and review the manuscript. These include Ray Bahm, Manny Landsman, Bud Lyon, Brad O'Mara, John Schaefer, Rob Wills, and Dave Miller. Input from the systems engineering group at Mobil Solar Energy Corporation was also of great help.

Andy Krantz and Paul Maycock deserve special appreciation for their encouragement and support of my work, as do many others, including Denis Hayes, Bob Hammond, and Bill Yerkes.

A book is a painfully slow way to communicate the progress and potential of a technology that is advancing as rapidly as photovoltaics. An electronic teletype continually flashing up-to-the-minute news of the most recent developments would seem more appropriate. But a book is the medium we have chosen and I owe an enormous debt of gratitude to my collaborator William Scheller for help in research and for translating many difficult technical concepts into readable prose.

My editors at Rodale Press—Joe Carter, Ray Wolf, and Larry McClung—also deserve special notice for their unfailing support and contribution. In addition, Kay Fissinger from ARCO Solar and Dot Bergin from Mobil Solar were especially helpful in providing photographs and technical information.

Finally, my appreciation to my family for the many late nights, early mornings, lost weekends, and missed holidays and vacations they have invested in this book.

Introduction

This book is about the solar generation of electricity and the ways in which it can be put to use in the households of the 1980s and beyond. The technology of solar electricity is called *photovoltaics*. Like so many of the new technological innovations that have been changing our lives over the past 20 years, it is based upon the remarkable properties of silicon and related semiconductor elements. The family of semiconductor devices is vast and its roster of uses even vaster—all because of an elemental ability to regulate the flow of electrical currents and, in the case of photovoltaics, to actually transform light energy into electricity.

The chapters that follow will explain how the photovoltaic principle works to transform sunlight directly into electricity, what sort of hardware does the job, how the source is hitched to the load, and how to go about designing a system appropriate to a specific need. But before we immerse ourselves in the details of this fascinating technology, we need to pause and consider how truly revolutionary it is. Once we begin to use photovoltaic devices and grow accustomed to their presence in our lives, we will be tempted to take them for granted as we do so many other wonderful inventions. But before that happens, we should reflect on how marvelous it is to be able to get electricity directly from the sun by means of noiseless, lightweight, solid-state solar modules—with no burning of fuel and no turning of dynamos or turbines.

The ability to accomplish this miracle signals what is perhaps the most important development ever in the history of energy production and distribution. It comes upon us at a time when conventional systems of generation and delivery are plagued by unpredictable shortages followed by temporary surpluses, incredible cost overruns, and highly unstable price structures. The terms *energy efficiency* and *cost-effectiveness* have come to dominate the vocabularies of consumers and producers alike. In this time of frustration and change, what sort of world will a technology as radically different as photovoltaics help create? Perhaps we can best approach this question by asking ourselves what sort of world we would like to have, since, to a greater extent than ever before, we now have clear options from which to choose.

Ever since the energy debate began in earnest, various factions have been presenting conflicting versions of the not-so-distant future. The contrasting images and scenarios most commonly described are those of centralization versus decentralization; "hard" versus "soft" energy technologies; a growth economy versus a steady-state economy; and traditional industrialism versus postindustrial society.

How will the widespread use of photovoltaics influence the outcome of these debates? In the dispute over whether or not energy production should be centralized (a situation best represented by large-scale fossil fuel or nuclear power plants feeding complex power grids), the inherent flexibility of solar electricity adds a newfound capability and credibility to those who favor decentralization. This is seen as a tremendous boon by energy analyst Amory Lovins, whose book *Brittle Power* (Brick House Publishing Co., 1982) is an indictment of the inflexibility and vulnerability that he claims are built into our present energy infrastructure. No doubt about it—photovoltaic cells are the great decentralizer, enabling anyone with an appropriately sized array to generate power on the very site where it will be used. But, as we will see, this same flexibility also makes photovoltaics well suited to supply large blocks of power to the central utility distribution grids and to serve in the intermediate range as a power source for remote villages, factories, and housing developments.

The "hard" versus "soft" energy debate centers on the notion that there are two basic ways of producing power. The first utilizes capital- and resource-intensive technologies and also contributes greatly to environmental pollution; the second makes use of far more manageable and forgiving technologies and is based on the use of renewable resources. While it is difficult to imagine a more sophisticated, up-to-date technology than photovoltaics, the fact is that solar electricity represents the "soft" approach in the best sense of the word. It can be used on a limited, local scale, without multi-billion-dollar investments, 15-year start-up timetables, and elaborate security precautions. It makes no further use of the earth's resources beyond the initial investment in manufacturing cells and auxiliary apparatus. Moreover, it creates no polluting by-products because nothing is either created or used up during its operation.

The effect of photovoltaics on the world's economy may well prove to be nothing short of spectacular. The major objection to sustained economic growth by critics favoring a no-growth, "steady-state" economy has been the enormous cost of perpetual growth in terms of pollution and resource depletion. Advocates of perpetual growth argue, in return, that unless we all make a concerted effort at growing richer, many of us will grow poorer—particularly those who are poor to begin with. Photovoltaics offers hope for a way out of this dilemma. This is one technology that can help fuel economic advancement, especially in developing countries, without forcing people to blindly accept belching smokestacks, oil spills, acid rain, and radioactive waste as the price of a higher per capita income.

Finally, we face the question of whether we are to live in an industrial or a postindustrial society. To some extent the distinction is fanciful, since humans are an industrious lot and will always pursue some form of industry. But the term *postindustrial* has a special meaning. It is used to describe an economic order in which high technology and information-oriented service industries replace a declining, older, manufacturing-based mass economy. Photovoltaics is an excellent example of the type of high-technology enterprise that will help bring about this change. In photovoltaics we have an infinitely flexible, multioption technology that can aid the movement away from a rigid, highly centralized, single-option infrastructure.

Whether future society is industrial or postindustrial, people will continue to want manufactured goods. If the sun's power can be harnessed to produce electricity cheaply and efficiently, those things can be produced with an energy support structure less elaborate than the one that currently serves industrial economies. This is especially good news for many parts of the Third World, where geographic and economic realities preclude old-fashioned industrial buildup. In such places, photovoltaics rapidly makes community-scale manufacturing enterprises possible.

In contemplating our own postindustrial future, we tend to forget that sizable portions of the human community have not yet experienced the industrial revolution firsthand. The great boon of solar electricity is that it can help emerging societies bypass the two centuries of smokestacks and slag heaps, along with the excessive centralization of wealth and energy resources, that the established industrial societies have had to endure.

Ultimately, the harnessing of electricity from the sun can help both the rich and the heretofore poor nations achieve an industrial flexibility that delivers the right type and amount of energy to the places where it is needed, without waste or excess. Obviously, photovoltaics alone will not solve all the world's problems, but this revolutionary technology can serve as an important step in the right direction.

The Solar Electric House begins with a general description of this exciting technology, then proceeds to show how it is relevant to the concerns of the homeowner. The bulk of the book is devoted to advice on how to design, install, and maintain a residential photovoltaic system.

Chapter 1 provides background information. There you will be introduced to the quantum theory of light and the basic physical principles involved in the photovoltaic reaction. You will learn how photovoltaic cells and modules are currently produced and will also become familiar with some of the new materials and processes that promise to make solar cells increasingly affordable in the future.

Chapter 2 explains the basic photovoltaic system options available to the homeowner and explores the economic considerations involved in making a decision for one or other of these options. Chapter 3 describes the way in which photovoltaic modules are combined into arrays and the methods of determining the optimum size of an array. Chapters 4 and 5 supplement that discussion with detailed information concerning the various components needed to complete a particular photovoltaic system.

Chapter 6 moves beyond a focus on types of systems and components to provide useful guidelines that will help you determine just how much electricity you really need. The importance of energy conservation through the use of efficient appliances is stressed, and different methods of load management are explored.

Chapters 7 and 8 lead you step by step through the process of designing a specific photovoltaic system. In chapter 7, the focus is on utility-interactive systems—those that work in tandem with the utility grid. In chapter 8, the focus shifts to stand-alone systems—systems that are independent of the grid and use batteries to store site-produced power.

Chapter 9 covers the installation of photovoltaic arrays and balance-of-system components such as inverters, batteries, and auxiliary generators, while chapter 10 explains how to wire and properly maintain a photovoltaic system. Finally, in the back of the book you will find a bibliography and a list of helpful addresses.

Primarily, this book is meant to serve as a practical guide for the design, installation, and maintenance of photovoltaic systems. For this reason, a lot of detailed information is provided, including specific examples taken from photovoltaic systems designed by my firm in various parts of the country. But even if you have not yet made a definite decision to install a photovoltaic system in your home, this book can be of value to you. Not only will you find much to satisfy your curiosity and extend your knowledge of the subject, you will also become aware of the many factors involved in intelligent system design. More important, you will learn which questions need to be asked and answered in order for you to make an intelligent decision concerning the possible installation of a solar electric system in your home.

When you first approach the subject, you may feel that you are entering a strange and mysterious realm, one that should perhaps be left to scientific experts and engineers to explore. However, once you become familiar with the terms, concepts, and physical principles involved, that initial feeling of strangeness will change into excitement and a fascination with the details of a rapidly developing technology. Most likely, that fascination will lead you to want to know specifically how you can apply this technology to your residential power needs, and that is what this book is really all about.

As final editing began on this manuscript, news flashed around the world of the nuclear meltdown at Chernobyl, and I was again reminded of how vitally important it is for this world that all people have a safe and sustainable supply of energy. Photovoltaics will play a major role in this new energy future, and it is my hope that this book will in some small way help to hasten this transition.

CHAPTER ONE

THE HISTORY AND DEVELOPMENT OF PHOTOVOLTAICS

On a bright spring morning in early May of 1981, a new type of house and a new era in homebuilding were heralded at a dedication that took place in the suburbs of Boston, Massachusetts.

Normally, the dedication of a new house is a celebration of completion. Construction work is finally done, and the building is ready to be occupied, lighted, heated, cooled, and enjoyed. Comfortable shelter and a certain measure of aesthetic satisfaction are usually all that is expected from this point on. This particular occasion, however, involved more than the usual laying down of tools, because the event being marked was the bringing to life of a house that is a wonderful new machine for lighting, heating, and living. It was the world's first energy-independent solar electric home.

The Carlisle house, as it has come to be called, appears on three sides to be no different from most other contemporary houses built with an eye toward energy conservation. In designing this house, I was frugal with windows on the north, east, and west elevations; windows placed in those locations are triple glazed for better insulation value. The earth is banked higher around these sides, and the outer tip of the gently sloping north roof almost reaches this raised ground.

Since lots of south-facing glass is a common design feature in energy-efficient houses, this is what you might expect to see on the Carlisle house. But as you face the south side of this house, you hardly notice the double-glazed windows and doors, although they do cover a sizable area. What is most remarkable is that the entire south roof is covered with photovoltaic (PV) modules made up of bright cells that make the whole surface look like a reflection of blue sky on still water. The only interruption of this expanse is an array of the more familiar solar thermal collectors used for domestic water heating.

The photovoltaic array on this house consists of 126 PV modules made up of 72 individual solar cells per module. Each module produces

PHOTO 1-1
This view of the Carlisle house from the south shows roof-mounted PV array and solar domestic hot water collectors.

about 14 volts of electricity at 4.2 amperes, or amps, and has a rated peak power output of 58 watts. (There will be a more complete discussion of these terms later.) These 9,072 cells provide virtually all of the home's annual electrical needs. The system's peak power output of about 7.3 kilowatts assures that the residents of this house can enjoy all the amenities of modern life. Inside are television, stereo, frost-free refrigerator/freezer, electric range and oven, clothes washer and dryer, ample lighting, outlets for tools and appliances, and even a whirlpool tub. Nor is there any compromise in climate conditioning. A high-efficiency, dual-compressor heat pump delivers heating or cooling as it is needed.

The exterior of the Carlisle house signals to the passing observer the energy-efficient nature of the structure. Less obvious but no less important are the many interior design features that contribute to the house's efficiency. Beneath the tile floors is a thermal mass of poured concrete, shaded from the summer sun by the overhanging roof yet open to the lower-angle winter rays. Warmed by the sun in winter, the concrete and tile radiate heat back into the rooms on cold nights. The great brick fireplace in the living room—situated opposite floor-to-ceiling, double-glazed doors—acts as additional thermal mass.

Warm and cool air circulate easily in this house, thanks to its balcony and cathedral ceiling. Clerestory windows can be opened for summer ventilation. The building envelope is constructed to superinsulation standards with double-stud R-40 walls and R-60 ceilings. All of these features add up to impeccable energy credentials for a modern home. But perhaps the most impressive credentials are provided by the numbers on a small meter mounted on the wall in the dining room.

The meter tells when the photovoltaic system is producing electricity and how much (in kilowatts) it is producing. When the sun doesn't

PHOTO 1-2
The living room of the Carlisle house.

shine, or when the home's electricity use outpaces the PV contribution, it indicates the amount of power being drawn from the local utility. Unlike some solar electric homes, the Carlisle house has no provision for electrical storage. Instead, it was designed to draw power from the utility when necessary and feed it back to the utility when the solar cells are turning out more power than the house can use. When that happens, a utility kilowatt-hour meter in the equipment room actually runs backward.

The financial advantages of this house can best be appreciated at the end of a month or a whole year, when the energy bills and credits are tallied. For the people who live in the Carlisle house, a reasonable program of energy conservation could easily bring the tally to an even draw—and maybe even put them into the black.

But the motive for interconnecting a photovoltaic system with the power grid is not necessarily to make a profit from selling electricity and certainly not to put the utility out of business. In a grid-connected system, the utility becomes a partner of sorts with the homeowner. It accepts the surplus electricity when the PV system's output exceeds the house load and provides power to the home when the PV output is less than the load. The attraction of such an arrangement is not that it makes us totally energy independent but that it enables us to control the production of much of the energy that we consume, while helping to make clean, abundant electrical power available to the larger society.

Though now privately owned, the Carlisle house was designed by my firm, Solar Design Associates, as a demonstration of residential photovoltaics and was constructed with the cooperation of the Massachusetts Institute of Technology and the U.S. Department of Energy (DOE). But the days in which solar electric homebuilding remained strictly within the province of scientific experimentation and government study are behind us. Already, PV systems are being installed by electricians rather than engineers, and the cost of their components is falling rapidly. Energy experts predict that by the end of this century, solar cells will be no less common than microwave ovens in new American homes. If you're inclined to be skeptical of this, just remember that a good pocket calculator that cost $400 a mere 15 years ago sells for a fraction of that amount today. As the technology continues to be perfected and the volume of production expands, we should witness a similar drop in the cost of PV systems.

A History of Photovoltaics

Cell—the individual part of a solar module that converts sunlight into electricity

Module—a sealed unit containing photovoltaic cells connected either in series or in parallel

Array—all the modules in a single photovoltaic system

What exactly do the photovoltaic modules and cells on the roof of the Carlisle house *do*—and how do they do it? For the answer, we must turn to what was, until recently, a curious backwater of electrochemical experimentation and to the strange and surprising world of particle physics. Like so many other important scientific ideas and principles, the idea of converting light directly into electricity—which is the basic meaning of the term *photovoltaic*—originated long before its practical applications were discovered.

The science of photovoltaics began with the pioneering work of the French physicist Alexandre Edmond Becquerel. He was the son of Antoine César Becquerel, a prominent scientist who had done extensive research in the generation of electricity through chemical reactions. In 1838, at the age of 19, Edmond began to assist his father in his duties at the Museum of Natural History in Paris.

During the following year, while conducting an electrochemical experiment involving two metal plates in a dilute acidic solution, he noticed that exposing the apparatus to sunlight increased the electrical energy it produced. Edmond Becquerel published his observations on this phenomenon and is generally credited with its discovery. However, at that time there was no practical use for the generation of electrical currents from sunlight and the idea was placed on the back shelf of science libraries.

Even if Becquerel's primitive apparatus had been vastly more efficient, it still would have predated by many years the devices it might have powered. Unfortunately, by the time Thomas Edison's procession of inventions began to leave his laboratory in Menlo Park, New Jersey, engineers were convinced that electricity could best be generated by burning fuels to create heat and move turbines. Today we are learning otherwise, and Edmond Becquerel may yet be celebrated for first observing the principle that will turn the wheels of the twenty-first century.

The Selenium Solar Cell

It would not be quite accurate to say that photovoltaics was completely forgotten during the rest of the nineteenth and early twentieth centuries. Any practical use, however, had to await not only the popular use of electricity but also the development of solid rather than liquid PV devices. The first major step in this direction was made in 1873, when Willoughby Smith first observed the light-sensitivity of the element selenium. Smith, an electrical engineer for the Telegraph Construction Company in Great Britain, was testing various metals in search of one suitable for use in underwater telegraph cables. Working with bars of metallic selenium, Smith discovered that in the absence of light selenium has a fairly high resistance to electricity, but when exposed to light it rapidly loses its resistance and becomes highly conductive of electricity. Through the use of a variety of filters, he discovered that selenium's ability to conduct electricity increases in direct proportion to the intensity of the light present.

This was the first observation of the *photoelectric effect*—that is, light causing the altering of the position of electrons within atoms—in a solid, and it led to experimentation and speculation regarding possible uses for a selenium solar cell. Such a cell was created in 1883 by Charles Fritts. However, despite the enthusiasm of its inventor and such farsighted supporters as the German engineer Werner Siemens, industrial and commercial acceptance was not forthcoming.

The stillbirth of selenium cells was not simply a matter of our forebears' lack of imagination. Selenium is a relatively uncommon element (it ranks 69th in order of abundance on earth), and the cells were expensive to produce. Also, selenium cells have never been capable of converting more than 1 or 2 percent of available sunlight into electricity. When we consider that even silicon cells were not taken seriously until their efficiency began to approach 10 percent, the reasons for selenium's failure become obvious.

Selenium—a relatively uncommon element whose sensitivity to light led to its use in the creation of the first solar cells

Still, selenium had—and continues to have—its place. The spectral sensitivity of selenium cells approaches that of the human eye; this has earned these cells an important place in light-measuring devices, particularly photographic light meters. For want of a better photovoltaic converter, researchers continued to tinker with selenium, as well as with cells of copper and cuprous oxide and various semiconductor–metal oxide compounds. But as late as the early 1950s, the efficiency ceiling of even the most advanced experimental cells was still unacceptably low.

The Silicon Solar Cell

In 1954, Bell Laboratories announced the developments that finally brought photovoltaics out of the laboratory and into the realm of practical application. For several years, researchers at Bell had been attempting to devise a more efficient solar cell for remote communication stations unconnected to utility grids. They were coming up against the old problems with selenium and were clearly in need of a different material. When one became available, it was a result not of their own experimentation but of work that was proceeding on a different Bell project—the refinement of techniques for processing silicon, a semiconductor material used in making then newly developed transistors and rectifiers (devices for converting alternating current into direct current).

Silicon—a nonmetallic element that, when properly treated, is quite sensitive to light and capable of transforming light into electricity

Silicon is a nonmetallic element related to carbon. Although it is second only to oxygen in abundance on earth (27.7 percent of the planet's crust is composed of silicon), it occurs in nature only in combined forms. Separated from the oxygen with which it forms silicon dioxide (silica) in rocks and soils, pure, solid silicon takes on a dark gray luster, which is in fact the color of all silicon solar cells beneath their antireflective coatings.

Word of silicon's photovoltaic properties traveled throughout the hallways at Bell. What had been discovered by the rectifier researchers was that silicon, when "doped," or treated with certain impurities, was quite sensitive to light and capable of generating a healthy voltage. Scientists already knew that silicon could turn light into electricity; the new findings centered on increased efficiency as it related to the careful introduction of impurities. Later in this chapter we will take a look at just what those impurities are and how they help solar cells function.

When the news reached the solar cell team at Bell, efforts were pooled and a joint research group began to develop a silicon solar cell. The result of the group's labors had an energy conversion efficiency of 6 percent—hardly impressive by today's standards, but a big leap from the 1 and 2 percent efficiencies of the selenium cells. "The device is not likely to

PHOTO 1-3

An early photovoltaic module from Bell Laboratories, circa 1955.

replace large-scale power plants—a 30,000-kilowatt battery would have to cover some 100 acres," reported *Scientific American* in its June 1954 issue. "But the company expects it to be useful as a small power source for such applications as rural telephone systems." Neither Bell nor *Scientific American* can be faulted for such conservatism in 1954. Still, foreseeing a photovoltaic future limited to rural telephones is roughly equivalent to observing the Wright brothers at Kitty Hawk and allowing that the airplane might someday be useful for skywriting.

Solar Cells in Space

Less than three years after Bell Laboratories announced its new silicon cell, the space race between the United States and the Soviet Union began in earnest. In the wake of Sputnik, the United States appropriated billions of dollars for its satellite and lunar exploration programs. The intensity of the competition was such that performance became a far more important criterion than cost-effectiveness, a state of affairs that could not help but benefit a promising yet still extremely expensive technology like photovoltaics. Engineers could not afford to ignore the sun's potential for meeting a spacecraft's energy needs. The sunlight available at orbital altitudes is unimpeded by atmospheric interference, and spacecraft constantly require power for communications, monitoring instruments, and controls.

Vanguard I, this nation's first satellite, incorporated a 108-cell photovoltaic system. But it proved to be too successful: with no way to turn off the power, the cells kept working and tied up a radio band long after the satellite had finished its useful life. The National Aeronautics and

PHOTO 1-4

All of the on-board power for this Skylab space station is provided by photovoltaic panels extending from the orbital workshop (lower right) and by panels attached at right angles to the Apollo telescope mount.

***Vanguard I*—the first satellite placed in orbit by the United States, it was powered by a photovoltaic system**

Space Administration (NASA) reverted to the use of batteries until a 15-watt PV array was built into Explorer VI in 1959. By that time, the military had recognized the value of solar-powered satellites, especially for communications. The U.S. Army's Project Courier, launched in October 1960, incorporated a 4-watt system. During the next four years, PV electricity output on individual American satellites increased by over a hundredfold.

As far back as 1945, science fiction writer Arthur C. Clarke envisioned communications relay satellites operating on solar energy. His predictions have long since come true, as photovoltaic cells have become an integral part of Telstar and its numerous successors. Missions such as Mariner, Voyager, Skylab, and the space shuttle have also relied on the rapidly improving technology of photovoltaics.

Solar Cells on Earth

Photovoltaics may have proved its value first in outer space, but practical applications of this technology on earth were quick to follow. True to the 1954 foresight of Bell Labs and *Scientific American*, many of these involved communications—although rural telephone systems were only part of the picture. Radio transmission in many remote areas is now powered, or is strengthened at repeater stations, by solar electricity. This is an attractive option not only in off-grid parts of developed countries but also in most of the Third World, where electric utilities often extend no further than major urban areas.

In such remote places, photovoltaics now powers lighting, irrigation, and even refrigeration—along with communications systems. One manufacturer has developed a line of special low-wattage refrigerators that are made to be used in conjunction with PV arrays to store perishable medicines in isolated corners of India and other developing nations. In the arid Arabian peninsula, PV projects are already providing electricity for water desalination, microwave repeaters, and cathodic protection of oil field equipment.

Third World and other overseas markets are, in fact, so receptive to solar electric technology that major photovoltaic manufacturers located in the United States have in recent years exported approximately 60 percent of their module output. Planners in developing countries are eager to enjoy the end benefits of industrialization while avoiding the traditionally difficult and expensive development phases. Photovoltaics is unique in its ability to help make this happen.

The world's first remote village to receive all of its electricity from photovoltaics, however, was not in India or sub-Saharan Africa but on the Papago Indian Reservation in Arizona. The settlement of Schuchuli (population 100) is served by a 3.5-peak-kilowatt (kW_p) array that powers simple direct-current appliances—both communal and individual—and provides basic lighting requirements for each home.

At Natural Bridges National Monument, located in a remote section of Utah, a 100-peak-kilowatt photovoltaic installation with ample battery storage capacity provides electricity for a visitors' center, six staff residences, maintenance buildings, and the water sanitation system. It replaced diesel generators that consumed 15,000 gallons of fuel per year.

PHOTO 1-5

This photovoltaic-powered desalination plant provides 3,000 gallons of purified drinking water a day to a desert village near Jeddah, Saudi Arabia.

On a vastly smaller scale, private owners of homes and farms located away from the power grid have also benefitted from the technological refinement and cost reductions that have characterized photovoltaic technology since its initial introduction in space applications. There are now between 15,000 and 20,000 small, remote homes in the United States receiving their electricity from photovoltaics, with over 4,000 in California alone. The economics of such remote installations are quite attractive when compared to the high-cost alternatives of fueling and maintaining a diesel generator or extending the utility grid to the site. (A detailed discussion and comparison of different energy strategies for remote sites is found in the latter part of chapter 2.)

PHOTO 1-6

This residence in Worthington, Massachusetts, is completely powered by photovoltaics.

Photovoltaic Power Plants

In addition to its very impressive track record in remote applications, photovoltaics technology is also proving itself a good working partner with the utility grid. In December of 1982, ARCO Solar, a subsidiary of Atlantic Richfield Company, turned on the world's first megawatt-size central-station PV power plant in the high desert near Hesperia, California. The system, constructed next to Southern California Edison Company's Lugo substation, took less than nine months from concept to completion and now feeds hundreds of megawatt-hours of electricity into the utility company's high-voltage distribution grid each month. The generating station operates unmanned under computer control and has experienced few interruptions in its normal operating schedule. This high degree of reliability contrasts sharply with the frequent shutdowns that occur at nuclear generating plants because of equipment failure, the changing of fuel rods, or the removal of waste from the site.

Quickly following on their success at Lugo, ARCO Solar entered into a cooperative agreement with Pacific Gas and Electric Company for the construction of a 7.5-megawatt central-station photovoltaic power plant at Carrisa Plain in San Luis Obispo County, California. The plant was completed in 1984 and enjoys the same problem-free history of rapid construction and reliable unmanned operation as its predecessor near Hesperia. While the capacity of these pioneer utility-scale PV projects is considerably smaller than that of conventional coal or nuclear central power stations, their success clearly demonstrates that there are no technical barriers preventing the construction and use of large-scale PV power plants.

In an era in which the 1,250-megawatt power plant is very quickly becoming a technology option of the past, perceptive utility executives are taking careful notice of the attributes of photovoltaics. With their short

PHOTO 1-7

This central-station photovoltaic power plant, constructed by Atlantic Richfield in the high desert near Hesperia, California, supplies power to Southern California Edison's distribution grid.

lead times, modularity, high reliability, and minimal environmental impact, PV systems offer a whole new set of options to the utility planner. Photovoltaic power plants can be constructed at the substation level in a matter of months, delivering reliable power close to the point of use and providing an extremely flexible, very low-risk, low-inertia investment. When comparisons are made with traditional generating plants, which are becoming increasingly difficult to site, take a decade or more to construct, require the commitment of very large amounts of capital for very long periods without return, and have intractable environmental problems, the case for central PV power stations becomes very compelling.

Photovoltaic-Powered Buildings

In recent years, utility-integrated applications of photovoltaics for buildings have also made significant advances. Since the completion of the Carlisle house in 1981, dozens of photovoltaic-powered buildings have been constructed and are now feeding power into the grids of utility companies across the United States. In the nation's capital, the south-facing roof of Georgetown University's new Intercultural Center generates 300 kilowatts of power, which is fed into the campus grid. In San Diego, the roof of an AM-PM Minimart generates 17 kilowatts and the surplus is sold to San Diego Gas and Electric Company. In New England, Massachusetts Electric Company engaged my firm, Solar Design Associates, to help plan the country's first photovoltaic-powered neighborhood community. The project involved the installation of 100 kilowatts of photovoltaics in dispersed roof-mounted systems on a single electrical feeder in Gardner, Massachusetts.

The proliferation of photovoltaic systems—large or small, private or communal, locally or federally financed, stand-alone or utility-interactive—points toward the day when photovoltaics will be a commonplace

PHOTO 1-8

This 17-kilowatt, rooftop photovoltaic array was designed and installed by ARCO Solar on an AM-PM Minimart in San Diego, California.

commodity as unremarkable as furnaces and as ubiquitious as water taps. But before that occurs, all of us have a great deal more to learn about the process by which sunlight is converted into electricity. The secret is part quantum physics and part screwdrivers and pliers. Discovering that secret and how it can be applied to residential use is what this book is all about.

Quantum Theory and the Nature of Light

Any understanding of the photovoltaic process must be based on an understanding of the nature of light. With the possible exception of gravity, no physical phenomenon has as great a constant effect upon us as light. Through photosynthesis, light primes the workings of the food chain. It also makes possible the perception of the colors, dimensions, and relative distances of objects. Yet it was not until early in this century that scientists discovered the proper vocabulary for describing the behavior of light and began to comprehend its physical properties.

Although the famed physicist Sir Isaac Newton argued three centuries ago for a corpuscular, or particularian, description of light, the prevailing model for explaining the movement of light from one place to another long remained that of waves. Light, it was believed, rippled through space in a wavelike movement similar to the movement created in still water when it is disturbed.

Physicists knew that this theory of light had its shortcomings and that there were light-related phenomena that it did not properly address. The wave model was not seriously challenged, however, until 1900, when the German scientist Max Planck first managed to formulate a mathematical description of thermal radiation from a perfect absorber, or blackbody (a theoretical object in which the intake and release of energy are in equilibrium). This led Planck to conceive of radiant energy as being quantized, or divided into particulate components—a return to Newton's corpuscular theory of light.

Quantum—a minute unit of energy

Photon—a quantum of radiant energy, or light

Albert Einstein carried Planck's work further and in 1905 proposed that light is made up of individual parcels of energy. On the submicroscopic level, he explained, light is both emitted and absorbed in discrete units called *quanta* (singular, *quantum*). In the language of quantum physics, which arose from Planck's and Einstein's work, a quantum is a minute energy packet of electromagnetic radiation. In 1926 the term *photon,* derived from the Greek word *photos* (light), was first used to describe those quanta that make up a beam of light. Although certain aspects of the behavior of light can still be described in terms of wave motion, the concept of the photon is extremely useful for explaining basic phenomena such as the photoelectric effect. (Actually, as you can see in the following discussion of solar radiation, the two models can be synthesized.)

In explaining photoelectric reactions (the term *photovoltaic* refers specifically to light-activated systems designed to produce an electrical current, whereas *photoelectric* refers simply to the excitation of electrons by light), Einstein demonstrated that light striking an electrically sensitive surface acts as if all of its energy is divided among the quanta, or photons, of which it consists. The work accomplished by this bombardment of photons depends upon the nature and adaptation of the material. Later, we will examine the nature of silicon, the material found in most solar cells, but first let us consider the properties of sunlight, the driving force of photovoltaics.

Solar Radiation and Photovoltaics

Virtually all of the energy available on earth, from wind and biomass to coal and petroleum, originates within the thermonuclear fires of the sun. Solar energy comes to earth in the form of waves of radiation—waves of photons possessing varying amounts of energy and traveling at a uniform 186,000 miles per second. The length of these waves depends upon the extent of that energy: the more energy a group of photons possesses, the shorter the wavelength. (Similarly, the shorter the wave, the greater the energy content of its photons.) We have developed a means of identifying the varying solar wavelengths, based upon the range of visible light as well as those upper and lower reaches that are imperceptible to the human eye. The entire continuum is called the *solar spectrum.*

***Solar spectrum*—the entire continuum of solar light waves of varying lengths**

At the lower end of the solar spectrum are the low-frequency, invisible wavelengths called *infrared,* which make up nearly half of the sunlight reaching earth. Next comes the visible spectrum, with wavelengths growing progressively shorter from red to violet. Once the infrared and visible wavelengths are accounted for, only 5 to 7 percent of the solar spectrum remains. This is taken up by the wavelengths beyond violet, or *ultraviolet* radiation. Their frequency—the distance from wave crest to wave crest—is the shortest of all, which allows us to describe their agglomerate photons as possessing a relatively high amount of energy.

***Infrared*—low-frequency solar light waves invisible to the human eye**

***Ultraviolet*—high-frequency solar light waves invisible to the human eye**

Not all of the wavelengths of solar radiation are readily convertible to electricity within a photovoltaic cell. Those wavelengths that contain the highest-energy photons are most useful. This amounts to roughly three-quarters of the solar spectrum. The remaining quarter, comprising most of the infrared zone, has no part in the process, even though some infrared light is absorbed by silicon atoms, as can be seen in figure 1-1. Even the radiation that occupies the ideal spectrum of wavelengths is not available in its entirety to photovoltaic cells. This is because of the filtering and reflective qualities of the earth's atmosphere and the constantly changing orientation of the sun.

Earlier we noted that photovoltaic cells are an ideal power source for spacecraft because of the absence of atmospheric interference at orbital altitudes. At ground level, it's a different story. Approximately 10 to 15 percent of solar radiation is absorbed by ozone, water vapor, or carbon dioxide in the atmosphere and thus never reaches the earth's surface. Another 30 to 35 percent doesn't even make it that far; it is reflected off the atmosphere and accounts for the phenomenon called *earthshine.*

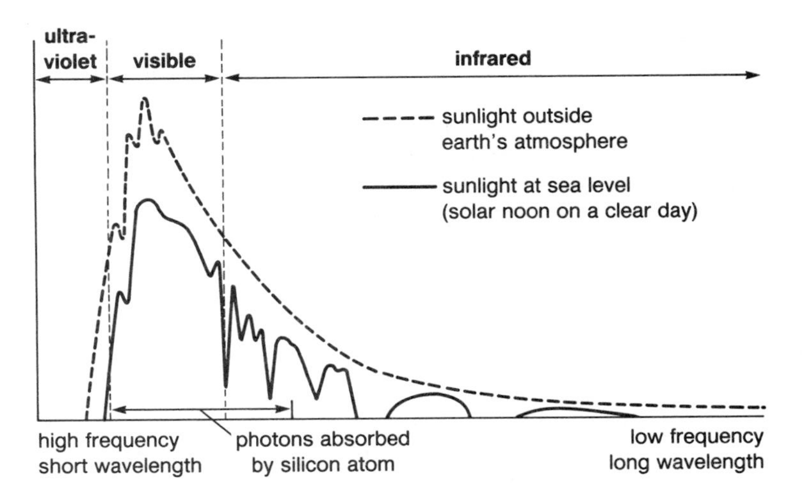

FIGURE 1-1

The solar spectrum, showing the difference between the levels of radiation found outside the earth's atmosphere and those found at ground level. Note that only the higher-energy photons are useful to photovoltaics.

The output of a photovoltaic module or array is directly proportional to the amount of sunlight it receives, in terms of both intensity and duration. Seasonal and daily variations in available sunlight are a function of the earth's rotation and its angle of tilt toward the sun. In simplest terms, this explains why sunlight becomes more abundant toward noon and tapers off in the afternoon and why it falls upon a given area with less intensity during the winter. Even during the summer months, when solar radiation is most direct, there are optimum angles at which collecting devices—photovoltaic or thermal—ought to be placed. These vary according to latitude and are discussed in chapter 3.

Insolation, which is shorthand for *incident solar radiation,* is a means of measuring the amount of solar radiation available at a specific place and time. Tables containing insolation data for all major areas within the United States provide a quick way to see the essential differences between places like Boston and Phoenix. (Insolation data for a selected group of U.S. cities is provided in chapter 3. Sources with much more complete information are listed in the bibliography.)

Insolation data is very useful in the design of photovoltaic systems. But such data alone cannot answer the question of how cost-effective such a system may be in a particular location. Many variables must be taken into account in making that decision, as you will see in our later discussion of the economics of photovoltaics.

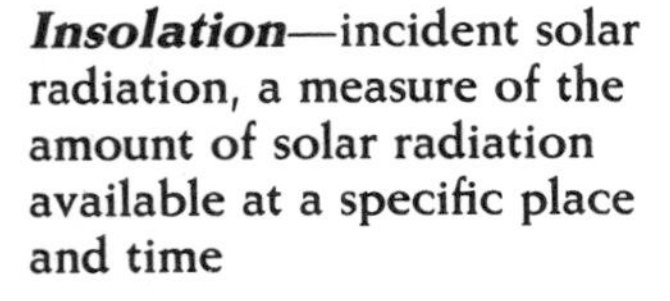

Insolation—incident solar radiation, a measure of the amount of solar radiation available at a specific place and time

Conductors, Insulators, and Semiconductors

Photovoltaic electrical systems, like modern computers, rely upon the principles of semiconduction. *Semiconductor* is a word that 30 years ago had hardly left the laboratory, but today it turns up almost daily on the business pages of our newspapers. It is nevertheless a term that is used more often than it is understood. Since, in fact, a PV cell *is* a semiconductor, an explanation of solar electric technology must include a discussion of the basics of semiconduction.

Each of the elements in the periodic table is characterized by a different atomic structure. Gold and lead, for instance, differ not only in outward appearance but also in the makeup of their atoms. The nuclei of disparate atoms contain varying numbers of protons and neutrons and are orbited by successive layers, or shells, of electrons, whose numbers also vary from one element to another. The outer shell of orbiting electrons is called the *valence shell.* It is this part of the atom that affects the conduction of electricity and that concerns us here.

Valence shell—the outer shell of orbiting electrons; the part of an atom that affects the conduction of electricity

In terms of electrical conductivity, the elements may be divided into three basic groups: conductors, insulators, and semiconductors. Elements whose atoms have only one or two electrons in their valence shells are said to be metallic and are good *conductors* of electricity. The reason is that those outer electrons are easily disengaged from orbit around their nuclei and induced to create or contribute to an electrical flow. Electrons knocked free of their atoms occupy what is called the *conduction band.*

Another group of elements possesses an atomic structure in which six or seven electrons make up the valence shell. These electrons are more closely integrated into their atoms and cannot be so easily jarred into the conduction band. Because of their strong electrical resistance qualities, these elements are good *insulators.* Think of a live electric wire with a rubber-insulated outer covering. The easily loosened and free-traveling electrons in the copper wire are surging along in the form of an electrical current. But when they reach the atoms that make up the molecules of the rubber insulation, they meet resistance in the form of electrons that are far more difficult to dislodge. This is the phenomenon that, if everything is working properly, contains the electricity in the wire, keeping it from leaking out in undesired places.

Conductors—elements whose outer electrons are easily disengaged from orbit around their nuclei, thus contributing to the flow of electricity

Insulators—elements whose electrons are closely integrated into their atoms and are not easily jarred loose, thus making them strongly resistant to electrical flow

Between conductors and insulators lies a third class of elements. If an atom has three, four, or five electrons in its valence shell, it is only marginally conductive of electricity, and the element to which it belongs is termed a *semiconductor.* The usefulness of semiconductors lies in the facility with which their ability to conduct an electrical current may be controlled, given the proper outside stimuli. Silicon, with its four valence-shell electrons, fits the requirements for semiconduction quite nicely.

Semiconductors—any of a number of elements, including silicon, whose electrical properties lie between those of conductors and insulators, making them only marginally conductive of electricity

Silicon and the Photovoltaic Reaction

What exactly is the nature of the photovoltaic reaction between sunlight and silicon? The answer to this question requires a correlation of the quantum theory of light with our understanding of the semiconductive properties of silicon. When Albert Einstein published his 1905 paper on

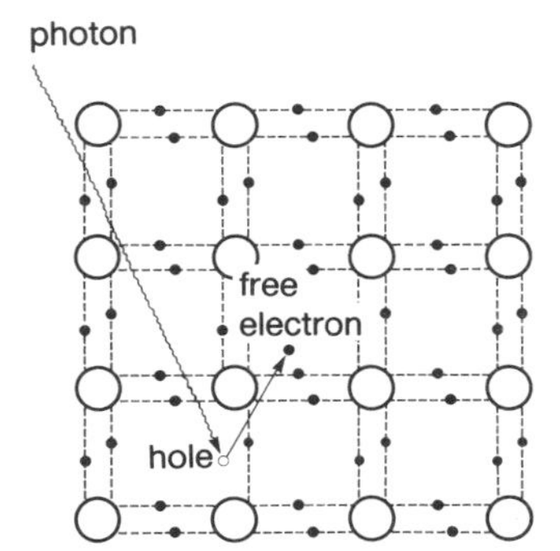

FIGURE 1-2

A diagram of the lattice structure of single-crystal silicon. When photons in light of sufficient energy strike an atom of silicon, an electron is knocked free to become part of the electric current produced.

the photoelectric effect, he postulated that photons are able to penetrate not only into the surfaces of matter but also into atoms. These photon/atom collisions, he stated, could cause an electron to be knocked loose from an atom's valence shell (see figure 1-2). The success of this reaction depends upon the atomic structure of the material so exposed and also upon the energy inherent in the photons.

When a photon of sufficient energy strikes an atom of silicon, it pushes an electron from the valence shell into the conduction band, where it is free to become part of an electrical current. (This seems a very benign way to harness the power of the atom when we consider the consequences of the bombardment of uranium atoms by liberated electrons in the reaction called *nuclear fission*.) But free electrons do not a useful current make—the basic photon/electron reaction must be induced to work in a certain way.

Consider the submicroscopic situation in a piece of silicon exposed to light. The atoms are absorbing photons, and electrons are being displaced. (Photon energy not converted into electricity is given off as heat.) The empty places in the valence shells, where electrons used to be, are called *holes*. These are appearing as rapidly as the electrons can vacate them, but they are being filled just as rapidly by dislocated electrons from neighboring atoms. The whole frantic exchange begins to resemble a game of musical chairs in which the players have forgotten to subtract the necessary chair. There's plenty of activity, but no headway is being made.

The problem remains: How can we obtain a flow of useful electrical energy from what appears to be only a self-contained, solid-state game of electronic musical chairs? It is not enough that the electrons are moving rapidly about within a solar cell. They must be forced to leave. In terms of our game analogy, some of the chairs have to be removed to keep the players moving.

Doping Silicon

Silicon doping—**a process whereby small quantities of boron and phosphorus are added to pure silicon, creating the imbalance within the silicon atoms that allows the photovoltaic reaction to occur**

In order to get the electrons flowing out of a cell, photovoltaic engineers resort to a technique called *doping,* which involves the introduction of special "impurities" into the cell's silicon. Remember that atoms of the elements in the semiconductor category have three, four, or five electrons in their valence shells. Silicon has four. The element boron has three, and phosphorus has five. Boron and phosphorus are the *dopants* used in manufacturing silicon solar cells. When added to pure silicon in minute quantities, they create the imbalance within the silicon atoms necessary to make photovoltaics work.

Actually, there are two imbalances working together during a photovoltaic reaction. When silicon is doped with boron, its atomic structure is affected by the electron deficiency that the boron introduces. Instead of a regular series of pure silicon atoms with four electrons in their valence shells, the crystal structure of boron-doped silicon reveals an occasional extra hole, or place where an electron should be. Since electrons carry a negative charge, this deficiency results in an overall *positive* charge for the doped crystal. For this reason it is referred to as *p-silicon.*

Doping silicon with phosphorus produces the opposite effect. With its five-electron valence shell, phosphorus creates in the silicon an abundance of electrons rather than holes. The result is called *n-silicon,* because it carries a *negative* charge.

P-silicon—silicon doped with boron to give it a positive electrical charge

N-silicon—silicon doped with phosphorus to give it a negative electrical charge

A silicon photovoltaic cell consists of separate layers of boron- and phosphorus-doped silicon. The analogy immediately suggested is that of a sandwich, although the n-silicon component is so much thinner than its positively charged counterpart that a better comparison might be made with icing on a cake. The thickness of an entire cell is generally no more than a hundredth of an inch, with the phosphorus-doped n-silicon accounting for only a few microns at the top surface.

Now we have two layers of silicon, one with a surplus of electrons in its atomic structure and one with a deficiency. It is the phosphorus-doped, electron-rich side of the cell that faces the sunlight, but both sides respond to the bombardment of photons upon their lopsided atoms. Electrons are driven into the conduction band, where they naturally begin to seek the spaces in the valence shells of atoms that have been vacated by other freed electrons. On the p-silicon side there are more holes to begin with, while in the n-silicon a shortage of holes prevails. What you would expect is that, given the activating effects of sunlight, the electrons and holes would pair with each other until a state of equilibrium is achieved. Fortunately, that is not what happens. If it were, the net electrical effect of all this subatomic restlessness would not toast a slice of bread.

The ultimate accrual of an electrical current in a photovoltaic cell is made possible by the presence of a static electrical charge that is created between the n-silicon and p-silicon layers. This charge region, which is only a few atoms thick, is called the *cell barrier* because of the resistance it offers to migrating electrons. The barrier is formed by an electrical reaction that accompanies the joining of the two layers and lasts for the life of the cell.

Cell barrier—a region of static electrical charge, only a few atoms thick, located at the point in a photovoltaic cell where the positive and negative layers meet

The usefulness of the cell barrier lies in the fact that it does not offer equal resistance to all of the photoelectrically agitated electrons in the cell. Remember that on the n-silicon side there are a lot of electrons and a shortage of holes. The free electrons there are thus moving at a lower rate of speed—or, if you prefer, with a lower energy level. But over on the p-silicon side, free electrons are far less abundant and can circulate with greater abandon. (Think of how much easier it is to move about in an open field, as opposed to a crowded room.) The livelier electrons from the p-silicon layer are thus better equipped to breach the static resistance of the cell barrier, while the n-silicon electrons are more frequently repulsed.

It doesn't take very long for the n-silicon layer to become top-heavy with electrons that, because of their accumulation, are beginning to build up the electrical pressure that we call voltage. Connect an external load between positive and negative silicon, and an electrical current will flow. To return to the musical chairs analogy, we have not only finally upset the balance between players and chairs but also have provided an exit for a steady file of losers.

The electrons thus channeled off from the n-silicon layer of a photovoltaic cell is eventually returned to the p-silicon layer, so that the process may continue. And continue it does, with no depletion of the silicon's

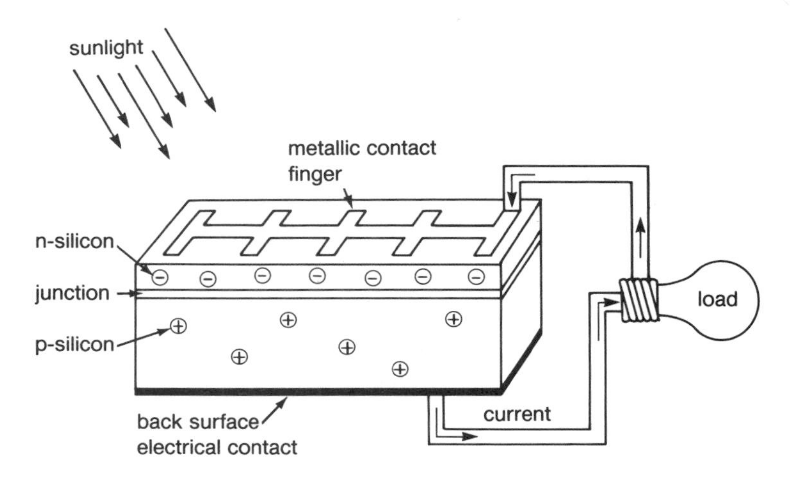

FIGURE 1-3

A diagram of a cross section of a photovoltaic cell. Freed electrons produced by millions of photon/atom collisions result in a flow of electrical current that can be harnessed through an external circuit to perform work.

mass and no essential changes in the cell's appearance or molecular structure. But in the meantime, we are free to intercept the flow of electrons and put it to use in powering a load.

In its simplest representation, then, this is the photovoltaic model: The energy present in sunlight (photons) frees the electrons in a cell of doped silicon. These freed electrons accumulate at a point from which they are allowed to flow out into an electrical circuit. The resultant electrical current can then perform work before it returns to the cell (see figure 1-3). No wonder that throughout the literature of photovoltaics, the word most frequently used to explain the process is *elegant*.

There is, of course, a limit to the amount of electricity that can be generated within a single solar cell. In conditions of bright sunlight, all silicon photovoltaic cells have an output of approximately 0.5 volt. This is a constant figure that does not vary with increases or decreases in cell area. (Remember, we are still talking about individual cells, not modules or arrays.)

Voltage—the pressure of electrons in an electrical circuit

Amperage—the measure of the current, or flow of electrons, in an electrical circuit

Wattage—the amount of power in an electrical circuit, determined by multiplying the voltage by the amperage

Voltage is a measure of electromotive force—the pressure of the electrons in a circuit—and is only one of the parameters that electricity comprises. Another is *amperage,* which represents the amount of current being generated and distributed. While voltage is a function of the cell's physical composition, amperage is affected not only by the area of the cell but also by the amount and intensity of light falling upon it. The third unit of electrical measurement is *wattage.* Wattage equals volts times amperes and is used to indicate the amount of power developed in an electrical circuit and, conversely, the electrical draw of a given load.

Increases in the voltage and amperage output of photovoltaic cells depend upon the way in which the cells are joined to each other in a PV *module* (one or more modules are assembled to form an *array*). In order to achieve higher voltage, the cells are linked *in series.* This means that the back contact of each cell is connected to the front contact of the one that follows it, as shown in figure 1-4. Two cells so connected would have a cumulative voltage of 1 volt, six would be rated at 3 volts, and so on. This

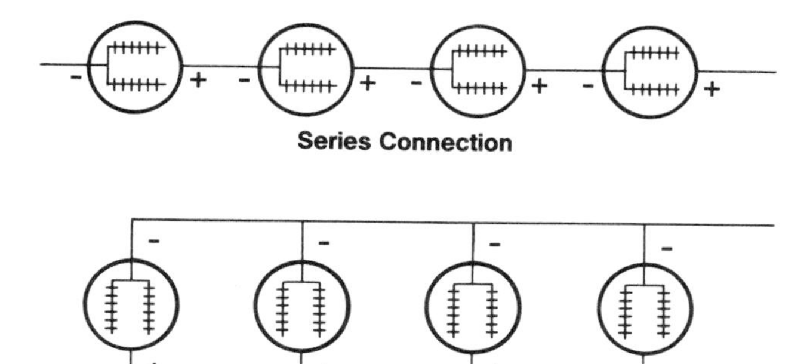

FIGURE 1-4

A schematic of solar cells connected in series. The back contact of each cell is connected to the front contact of the cell that follows it (positive to negative to positive).

FIGURE 1-5

A schematic of solar cells connected in parallel. The back contact of each cell is connected to the back contact of the cell that follows it and the front contact of each cell is connected to the front contact of the one following (positive to positive and negative to negative).

is the way in which cells are usually arranged in many commercial modules. The amperage, or current output, of such a module would nevertheless be the same as that of an individual cell. In order to boost amperage, the cells must be connected *in parallel*—that is, front to front and back to back, as shown in figure 1-5. Photovoltaic arrays commonly have modules wired both in series and in parallel. The proper combination of series and parallel connections in a PV array provides the necessary power output required to perform the work expected of it.

Photovoltaic-Grade Silicon

Several characteristics serve to differentiate ordinary silicon from the silicon in a photovoltaic cell. First, photovoltaic or other semiconductive reactions are accomplished most efficiently in a material having a crystalline or polycrystalline structure. In a crystal, all of the atoms and molecules are arranged in a definite, constantly repeated pattern. A polycrystalline sample is one in which several crystal patterns occur, with the lines of demarcation between them being known as *grain boundaries.* Most of what are commonly called crystals are actually polycrystalline structures; the few exceptions are naturally occurring gemstones and the carefully fabricated crystal semiconductors required by the electronics and computer industries.

Grain boundaries—lines of demarcation between different patterns found within a polycrystalline structure

In the past, scientists thought that photovoltaic performance fell off significantly when polycrystalline material was substituted for the single-crystal variety, but this now appears not to be the case. And since a perfect single crystal is a good deal more difficult and expensive to produce than one that exhibits grain boundaries, improved performance must be balanced against increased expense. Since the degree of incremental improvement in cell performance tapers off sharply as the single-crystal state is neared, the most attractive cost-efficiency ratios are currently achieved with silicon that is somewhat less crystallically perfect than that used by the semiconductor industry.

There is even a place in photovoltaics for *amorphous silicon*—that is, silicon whose atomic or molecular structure makes no pretense of regular-

Amorphous silicon—silicon whose molecular structure lacks regularity

ity. Such noncrystalline silicon responds photoelectrically to sunlight, although its efficiency is markedly lower and a larger cell area is required to produce the same amount of electricity as an array of crystalline or high-grade polycrystalline cells. Very thin films of amorphous silicon are currently used in solar-powered watches, calculators, and other small, low-wattage consumer products, and this approach holds considerable promise as a major photovoltaic technology in the future.

Since photovoltaic-grade, polycrystalline silicon requires less stringent manufacturing procedures than the single-crystal variety, you might think that producing an adequate supply would be relatively easy. Theoretically, it is. But since the computer industry and others that use semiconductors grew to maturity while solar electric technologies were still in their experimental stages, existing silicon-processing facilities are geared toward single-crystal production of semiconductor-grade silicon. Tremendous capital investment in new manufacturing facilities will be needed if polycrystalline silicon is to be produced in the quantity needed to create a cheap and abundant supply of solar-grade silicon. Fortunately, there is much evidence that indicates such investment is now under way.

Manufacturing Solar-Grade Silicon

The first step in putting together a silicon photovoltaic cell is the processing of raw silicon. Two objectives dominate this phase of the operation: first, the silicon must be virtually free of impurities, except for the de-

PHOTO 1-9
Round boules of single-crystal silicon prior to being sliced into wafers.

liberately added dopants; and second, it must be delivered in a crystalline or acceptable polycrystalline state. As noted earlier, crystallic perfection is not nearly as imperative in solar-grade silicon as it is in the silicon used by the electronics industry. Manufacturers of solar silicon also have more latitude when it comes to purity. Instead of the one part per billion impurity ceiling mandated for computer-chip silicon, impurities in the base PV material need only be held to one part per million. That may still seem impossibly finicky, but it translates into a considerable reduction in processing costs, which helps make photovoltaics affordable.

Single-Crystal Silicon

In the early days of solar cell manufacture, single-crystal technology was predominant. It was dependent largely upon the *Czochralski* method of growing, or "pulling," a perfect crystal. In this process, pieces of pure refined silicon are melted in a crucible. A seed crystal is then inserted into the liquid silicon and withdrawn at a slow, precise rate. Temperatures above the molten surface are controlled so that solidification proceeds according to an orderly pattern. This pattern is nothing more nor less than a repeat of the crystallic structure of the seed, resulting in one boule—a solid, cylindrical, single-crystal ingot ready for processing into solar cells.

Czochralski process—a procedure for producing large single crystals of silicon by slowly drawing a seed crystal from a crucible of melted silicon

This is the classic, first-generation method of producing solar-grade silicon, and it served demanding users such as NASA well. (All circular and many square photovoltaic cells manufactured today are cut from Czochralski ingots.) The quality and efficiency of this material are high, but it is in the next phase of processing that problems of practicality and expense are encountered. First, the cylinders have to be sliced into wafers. If this sounds simple, remember that a brittle ingot of silicon cuts more like a 2 × 4 than like a salami. As with a 2 × 4, each cut results in a kerf, equal to the width of the blade, in which the material is converted into sawdust. Now imagine what a disheartening sight a handful of sawdust would be if 2 × 4s cost $200 per pound. This is the great disadvantage of slicing wa-

Wafer—a slice of silicon or other semiconductor from which a photovoltaic cell is made

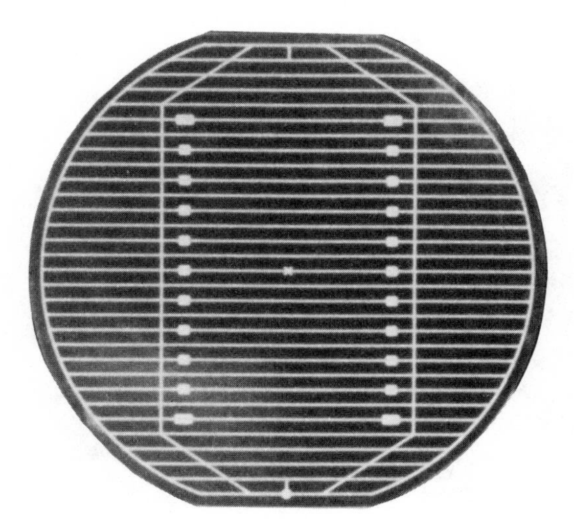

PHOTO 1-10

A finished single-crystal silicon cell manufactured by ARCO Solar from a round boule.

fers from silicon cylinders; each slice is so thin, and the cuts are so numerous, that up to one-half of the silicon ingot is wasted.

The search for a solution has led to ever-thinner saw blades and wire-cutting devices and the mechanization of labor-intensive hand-finishing steps. All of this has brought prices far below the $2,000-per-watt cost incurred for photovoltaics on the pioneer satellites of the 1950s. Still, some observers question whether, even with these improvements, this early crystal-pulling method will ever produce low-cost cells for the "mass market." Manufacturers who continue to cast their lot with Czochralski wafers argue that single-crystal cells yield greater electrical output early and late in the day and are more efficient than polycrystalline cells, especially in unfavorable weather conditions and extreme latitudes. But, as competitors of single-crystal photovoltaics steadily improve in efficiency, their lower production costs become ever more attractive.

Polycrystalline Silicon

There are two widely accepted ways of getting around the problem of sawing wafers from a single-crystal ingot. One involves reducing the expense of the ingot and hence the seriousness of the waste; the other completely eliminates the wafer-sawing process.

With the first option, the carefully pulled, ultrapure ingot of single-crystal silicon is replaced by a cast ingot with a polycrystalline structure. The expense of this process is lower because it doesn't require as pure a grade of silicon as feedstock and it is not as tedious as Czochralski crystal growth. The slicing methods are essentially the same, but you don't lose a small fortune every time the saw hits the ingot. It's more like cutting chipboard than teak. Several variations on this idea are currently in commercial production. One is Solarex Corporation's SEMIX process, which is responsible for the cells that power the Carlisle house.

PHOTO 1-11
A polycrystalline silicon ingot, a sliced wafer, and finished cells manufactured by Solarex Corporation.

The cast polycrystalline ingots are square or rectangular in shape, which is a noteworthy advantage over cylindrical silicon stock. When round cells are mounted in a module, there is a fair amount of dead space between them. This is obviously not true with cells cut from blocks. Since dead space generates no kilowatts, a photovoltaic module of a given size made with square cells will be more efficient than a module of the same area that incorporates round cells of comparable efficiency. Manufacturers of single-crystal photovoltaics have recently begun to machine their round boules into a square cross section before cell slicing to take advantage of the efficiency advantage square cells offer.

Boule—a cylindrical, single-crystal silicon ingot formed synthetically in a special furnace

Crystal Ribbons

Quite apart from the ingot technologies are those methods used to produce ribbons of silicon in single-cell thicknesses. For the past several years, the Massachusetts-based Mobil Solar Energy Corporation has been working with a technique called *edge-defined film-fed growth* (EFG) and marketing the results in its line of photovoltaic modules. In this process, a graphite die is inserted into a crucible of molten silicon. Capillary action draws the liquid up into the die, where it meets a seed-crystal ribbon that induces a continuity of crystal growth. When lengths of about 6 feet have been achieved, the hardened ribbons are cut free, although the alternative of reel storage has been explored. The ribbons are not saw-cut but are laser-cut to minimize waste when they are separated into cell-size segments. Viewed in operation, a row of the cauldrons and their emergent ribbons resembles a futuristic loom, turning out not cloth but solid-state electric power for the century to come.

EFG (edge-defined film-fed growth)—a technique for mass-producing crystal ribbons of solar-grade silicon

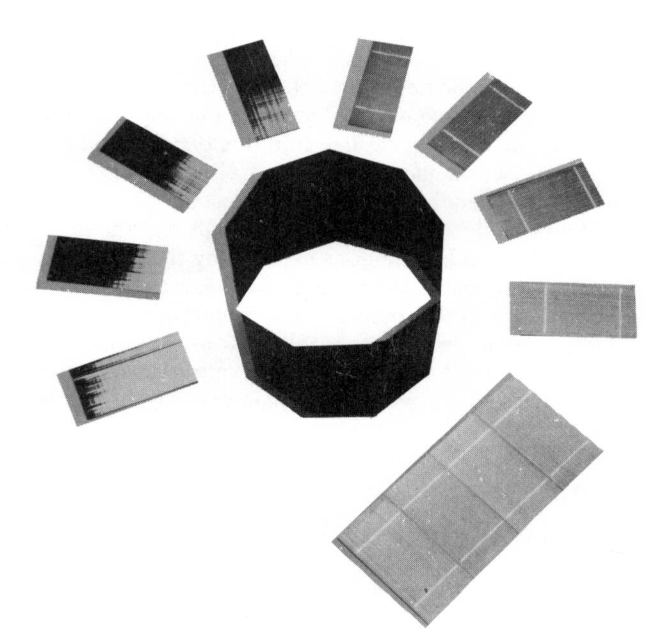

PHOTO 1-12

Mobil Solar Energy Corporation's silicon ribbon "nonagon," wafers, and finished cells.

Having eliminated the need for and the inherent waste of slicing silicon crystals to produce solar cells, engineers and researchers at Mobil Solar turned to address another of the basic problems standing in the way of mass production of low-cost solar cells: growth rate, or throughput. A round boule of Czochralski-grown crystal silicon grows very slowly, with a throughput rate measured in inches per hour. With the introduction of the single silicon ribbon of EFG material, with growth rates measured in inches per minute, Mobil Solar significantly increased the output of a crystal growth furnace.

To increase the rate of production of their silicon-ribbon material still further, Mobil Solar engineers developed a method of drawing molten silicon through a closed, nine-sided die to produce a nonagon tube. The tube, which looks like a shiny, silver-gray silicon stovepipe, is laser-cut directly into cell-size sections, or "blanks," that are then processed into finished ribbon solar cells. This process effectively increases Mobil Solar's production rate ninefold from each crystal growth furnace and is one example of the technological advances leading in the direction of the kind of mass production of photovoltaic cells that will lead to significant price reductions.

Dendritic Web Ribbon

Dendrites—**elongated starter crystals used in the formation of thin, weblike sheets of crystal silicon**

Westinghouse Corporation is experimenting with yet another ribbon growth technique called the *dendritic web method.* Like Mobil's edge-defined film-fed ribbon growth, this makes use of seed crystals brought into contact with molten silicon, but it dispenses with the die. Instead, two *dendrites*—elongated starter crystals—are lowered into the silicon. As they are drawn up, a sheet of crystal silicon forms between them. Although the chemistry, physics, and thermal conditions are different, the effect is not unlike the stretching of a soap solution across the rim of a child's bubble wand. But with the dendritic web, both the dendrites and the sheet between them continue to develop as the material is withdrawn, forming a flat crystal that is then cut free from the dendrites with a laser.

The Westinghouse dendritic web process is not without its disadvantages. Temperature regulation, for instance, is even more critical than with Mobil's edge-defined film-fed growth. But the major problem to date has been the width limitations imposed by the maximum 1-inch span between the dendrites. When the two seeds are placed farther apart, a third dendrite forms in the middle of the cooling silicon crystal film. As of this writing, no commercially available solar modules incorporate dendritic web cells, although the high crystallic integrity and potential photovoltaic efficiency associated with this technique justify further research.

From Wafer to Cell and from Cell to Module

We have looked at some of the ways photovoltaic-grade silicon can be produced. Let's turn now to the procedures involved in producing solar cells from this material and assembling these cells into modules. Fabricating the wafer and its p-n junction is only the first step. Subsequent steps include "metalization," or installing contacts for collecting and conducting electricity; applying an antireflective coating; connecting individual cells together to achieve the module's desired output voltage and current; and encapsulation, or sealing, of the modules to assure efficient electrical insulation, protection from the environment, and structural integrity.

There are many different cell contact designs, but each has the same function: to draw current efficiently from the cell's top surface without obstructing too much of its area. Since those portions of a cell that are covered by the metallic contact "fingers" are incapable of absorbing solar radiation, an inefficient contact design will reduce overall output of the cells and hence of the modules and complete arrays. In addition, the electrical resistance of both the contacts and the interface they create with the cells must be kept low in order to keep internal power losses to a minimum.

Since electricity must flow both *from* and *to* each cell, contacts must be provided at both the front (negative) and back (positive) surfaces. Interference with sunlight is not a prime consideration on the back of the cell, so the contact here is often a thin plating of metal over the full area of the cell. (In thin-film technologies, this plate can even be the substratum to which the semiconductor material is applied.) At the front, or n-silicon surface of the cell, contacts appear as a fine metal tracery in parallel, herringbone, pronged, or other configurations, all of which connects with one or more heavier contact bars running near or along the edge of the cell. These contacts are applied by a painting, printing, or plating method, depending on the pace of production and the methods favored by individual manufacturers, with the finished product resembling the printed circuitry found in any modern electronic device.

The metallic contacts on the front and back of each cell are also the points at which the series or parallel cell-to-cell interconnections are made. The connections require a linkage that is strong yet somewhat flexible to allow for thermal expansion and contraction. These interconnections made between cells should also be sufficiently redundant so that failure at one point does not jeopardize the entire module. The conductive elements used to join cells are called *interconnects.* A good deal of research has gone into their composition and design, resulting in an assortment of contemporary options that offer high conductivity, flexibility, and strength.

***Interconnects*—metallic conductors that electrically link the various photovoltaic cells within a single module**

Most of the basic 30- to 40-watt photovoltaic modules currently available on the market incorporate cells connected in series, while complete arrays may make use of both series and parallel connections between modules. However, many manufacturers are now building modules that combine series and parallel stringing of cells within the individual module, increasing reliability and providing more flexibility in module current and voltage output options.

After each cell is given its metallic contacts, the front surface is provided with an *antireflective coating.* The reason for this is simple: Silicon is highly reflective, and in its polished state throws back approximately 35

PHOTO 1-13
ARCO Solar's thin-film, amorphous silicon photovoltaic cells are available in many shapes and sizes to satisfy a variety of small power requirements.

percent of the light that strikes it. To counter this, manufacturers coat their cells with an extremely thin layer of a material such as silicon monoxide or titanium dioxide in much the same manner as photographic lenses are coated. It is the type of antireflective coating used that gives the different brands of solar modules their characteristic color, ranging from deep blue to nearly black. Some manufacturers are also now etching or texturing cell surfaces to cut down on reflection.

Antireflective coating—a thin layer of material added to silicon cells to prevent them from reflecting light

The reflection-resistant, interconnected strings of cells are now ready for final assembly into modules. Since they are the building blocks of photovoltaic arrays that may be expected to perform reliably for 20 years or more, the modules must be weatherproof, impact resistant, and capable of withstanding the environmental stresses, such as thermal cycling, wind, and snow loading, to which they will be exposed. Clearly, there is more involved in the making of a PV module than simply wiring together strings of cells and affixing them to a rigid surface.

Techniques of Module Assembly

As the module manufacturing industry has matured, the early practice of embedding cells in silicone rubber pottant has given way to more advanced and reliable multilayer lamination assembly techniques. The glass-superstrate module with a multilayer laminated encapsulation is now the industry standard.

Mobil Solar encapsulates strings of its rectangular ribbon cells in a seven-part, heat-sealed "sandwich" that has as its topmost surface a sheet of low-iron, highly transparent "water-white" tempered glass. (Except for its superior light transmission qualities, this material is similar to the glass used in car windows.) Next comes a sheet of ethylene vinyl acetate (EVA) for mechanical bonding and encapsulation, and then the cells. These are followed by a cushioning sheet of glass fibers, then another layer of EVA and one of Mylar, which provides the dielectric isolation important in high-voltage systems. Finally, the module's backing is applied in the form of a Tedlar/aluminum foil/polyester laminate, which seals the unit against moisture.

EVA (ethylene vinyl acetate)—a bonding material used in the manufacture of solar modules

Thus prepared, the module is placed in a lamination chamber, which performs the joint function of evacuating all air from the module and then heat-sealing the layers into an integral unit, with positive and negative ribbon conductors terminating at screw terminals for connection with adjacent modules in an array. The only steps remaining in production are module testing and enclosure of the laminated module in a lightweight frame of aluminum extrusion. Most major photovoltaic manufacturers now follow a similar procedure, with some variations in materials.

Photovoltaics Today and in the Future

The photovoltaics industry is destined to achieve tremendous growth and diversity over the next 15 years. The state of the art will progress like that of any dynamic new technology. While the underlying physical principles will remain the same, the means by which they will be translated into electric power will develop in hundreds of ways—some predictable, some yet unheard of. Thin-film silicon will play a larger role, as will semiconductor materials other than silicon. Concentrating collectors, which track the sun and focus its rays, will become more practical, as will thermal-photovoltaic hybrid systems that provide heat as well as electricity.

But if there is one point that deserves to be made again and again, it is that photovoltaics in the mid-1980s does not exist solely in the rarefied atmosphere of research and development—that it is not, as some would have it, an "exotic" energy technology longer on promise than on practicality. The techniques described in this chapter are being put to work in factories as well as in laboratories, and a wide variety of PV modules are already available as "off-the-shelf" components for purchase.

Photovoltaic manufacturers have largely concentrated upon creating a product that is modular in design, lending itself to incremental installation. The units available today are capable of filling a wide range of requirements: powering communications systems; electrifying ranch fencing;

pumping water; providing power for a home, a college campus, or a city. The chapters that follow describe system options for a wide variety of photovoltaic applications. As you will discover, different systems vary not only in size and level of complexity but also in the degree to which they actually depend on solar cells to provide total energy needs.

Part of the excitement of photovoltaics is the fact that the technology is in a very dynamic state. Manufacturers of solar cells and modules are not content to rest on past achievements but are constantly exploring new materials and new processes. The primary goal of research and development, both current and future, is to bring down the unit price of PV hardware. As prices fall, demand rises, adding momentum to the trend toward mass production of affordable PV equipment. When that occurs, millions of people will join the current thousands for whom the "solar electric house" is not a future dream but a present reality.

CHAPTER TWO

PHOTOVOLTAIC SYSTEM OPTIONS AND ECONOMICS

In chapter 1 we surveyed the historical development of photovoltaic (PV) technology. We also looked at the basic scientific principles underlying this technology in order to understand how PV devices function. Now, in this chapter, we are ready to explore various ways that photovoltaics can be put to work for our benefit. Our main object here is to consider different system options that can be assembled to serve particular needs.

One of the great advantages of photovoltaic generation of electricity is the modular approach it encourages. The degree of complexity of a particular system is closely related to the end-use demands that it is designed to meet. At the most basic level, a PV installation can be amazingly simple. As loads increase and versatility becomes more important, the system can be made as sophisticated as necessary to meet virtually any requirement. We will review a variety of system options, starting with the most basic and working up in order of size and complexity. But first we need to note the type of electricity generated by PV systems, whatever their size or type.

Types of Electricity

Electricity comes in two basic types: direct current (DC) and alternating current (AC). In a DC power source, the polarity and electrical pressure (voltage) remain constant—a car battery, for instance, has a positive terminal that always remains positive and a negative terminal that is consistently negative. With alternating current, however, polarity and voltage are constantly reversing. Examples of an AC power source are an automobile alternator (whose AC output is "rectified," or converted to direct current) and a generator of the type used by utility companies.

The output terminals in an AC generator rapidly and perpetually

Direct current—an electrical current in which electrons flow in only one direction

Alternating current—an electrical current in which flow direction is reversed 120 times (60 cycles) per second

change from plus to minus and back again 60 times per second. Since conventional current always flows from plus to minus, its direction is likewise reversing with the same regularity. The voltage reverses as well (from zero to a maximum level), but remains at an effectively constant average voltage (called the RMS voltage) because of the rapidity of the reversal.

Photovoltaic devices are capable of producing only DC electricity. If AC electricity is desired for a PV power system, a piece of electronic wizardry called an *inverter* is required to convert the DC output from the PV array into AC power. (Inverters will be discussed in detail in chapter 5.)

Basic Photovoltaic System Options

Photovoltaic systems take many forms. Some involve storage of current not immediately needed; others do not. Some are equipped with devices to regulate the flow of current; others have no such devices. Some use inverters to convert DC output into AC power; others are used only with equipment capable of operating on direct current. Some are linked to the utility grid and interact with it; others stand alone. As you can see, the options available are quite varied. The advantages and disadvantages of each system must be carefully considered in order to determine which one best suits a particular application.

DC Stand-Alone Systems

Load—the device or group of devices in a power system that consume the power; also the total amount of power required

The most basic scenario for the application of photovoltaic electricity to a load is a DC stand-alone (SA) system, a remote electrical system designed and dedicated for a single purpose, where the DC output of the PV array is delivered directly to the load with no storage and no control or regulation. This type of system will typically power a load that requires power only when light is available. Most PV-powered consumer products, such as calculators, use this approach.

One of the most common applications of this system configuration is the pumping of water with a DC variable-speed, variable-voltage motor. No electrical storage is provided. Instead, the product (in this case, water) provides the storage. The pump operates when electricity is supplied from the array, bringing the water above ground where it is stored for later use. This is a reliable, low-maintenance setup ideal for shallow wells (see figure 2-1). By contrast, batteries or clever control electronics are required to supply the high-amperage, inrush surge current needed to start the more powerful pumps needed for deep well pumping.

The next order of complexity in photovoltaic systems involves the use of a storage medium for array output not immediately used by the load. The configuration is still quite simple: The PV array's DC output goes directly to the load, with excess stored in a battery or bank of batteries. In the most basic version of this scheme, no voltage regulator is used. This means that the PV module or array of modules must be carefully sized to provide a self-regulating flow of current to the batteries. If this is

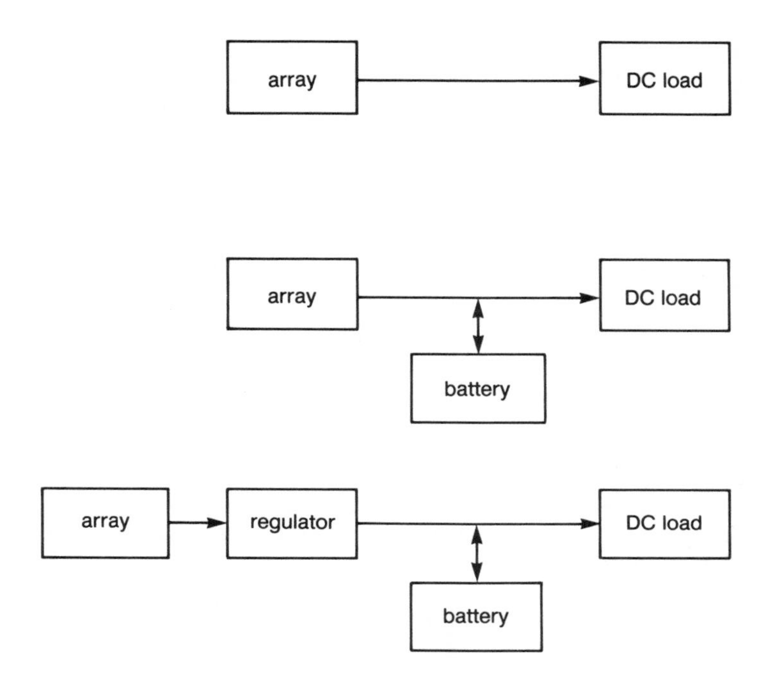

FIGURE 2-1 A schematic of the most basic DC stand-alone system in which the PV array is connected directly to the load with no storage provided.

FIGURE 2-2 A schematic of a basic DC stand-alone system in which the PV array is connected directly to load with storage provided.

FIGURE 2-3 A schematic of a basic DC stand-alone system in which the PV array is connected to the load and the battery bank via a voltage regulator.

not done, the batteries' state of charge must be monitored on a regular basis to prevent overcharging. In either case, the system must be checked frequently to ensure that the batteries are not being discharged too deeply. If this is the case, then the load must be disconnected manually to prevent further drain on the batteries. A typical application for this rudimentary system, depicted in figure 2-2, is a small vacation cabin in which DC lighting is the only significant load.

The next step in system sophistication is to install a voltage regulator between the array and the batteries to automatically control the flow of current to the batteries. In this way the proper state of charge can be maintained, prolonging battery life. The voltage regulator is an indispensable component in every photovoltaic system that is to be left unattended. Examples include remote communications relay stations, all but the most basic residential installations, and all other PV applications where daily monitoring of battery state of charge is not practical. Even with a regulator, a straight-DC system of the type shown in figure 2-3 would seldom be used in residential photovoltaic systems that have an array much larger than 500 peak watts (kW_p). The reason for this will become clear later in this discussion.

Basic AC/DC Systems

The next stage in complexity is a system that offers the AC option as well as a supply of DC electricity. This system, like the one just described, consists of the photovoltaic array, a voltage regulator, and storage batteries. The new element here is a DC-to-AC inverter. Note that we are

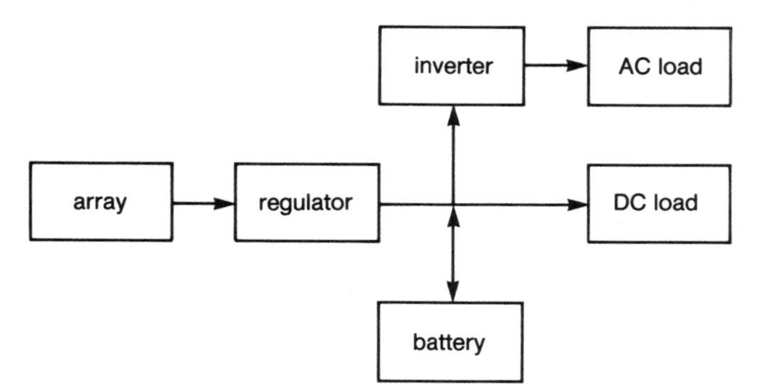

FIGURE 2-4

A schematic of an AC/DC stand-alone PV system equipped with a voltage regulator, battery storage, and a DC-to-AC inverter.

still talking about an SA system; the AC power is entirely site generated and typically will be part of a scenario calling for between 500 and 1,500 peak watts of array output. As shown in figure 2-4, the inverter is interposed between the array, regulator, batteries and the AC load. The DC load in this application is still served by a direct supply of current from the array and batteries.

With its ability to supply power to a wide variety of AC appliances, the system outlined in figure 2-4 represents a substantially more versatile option than the DC-only configurations. However, it still does not address the shortfalls between photovoltaic array output and load demand that might occur after a long period of low insolation and/or the discharge of the batteries to their maximum recommended depth. To deal with this problem when it arises, many owners of stand-alone PV systems opt to install an auxiliary engine-generator set, or "gen-set." This is simpler and less expensive than increasing the size of the PV array and battery storage to cover every extreme in weather and load conditions that potentially may occur. A stand-alone PV system that includes a generator is depicted in figure 2-5. Generators of the type shown are usually powered by propane or diesel fuel.

Here's how the auxiliary power source does its job. When the output of the photovoltaic array is insufficient and the batteries are approaching their maximum advisable depth of discharge, the generator is brought

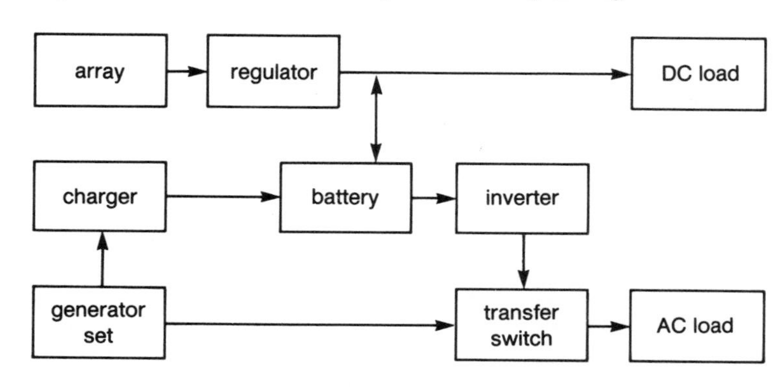

FIGURE 2-5

A schematic of an AC/DC stand-alone PV system with a voltage regulator, battery storage, an inverter, an auxiliary generator, and a battery charger.

into service. The power output of the generator, transferred to the batteries via an AC-to-DC battery charger, brings the storage bank back up to its optimum charge level while at the same time supplying DC loads. The AC side of the system, meanwhile, is served directly by the auxiliary generator's AC output, delivered through a transfer switch that disconnects the inverter and connects the generator output to the load distribution.

Keep in mind that in most systems of this type, the generator is considered only a backup device. It is used only occasionally, primarily to recharge batteries. The photovoltaic array is designed to provide the bulk of the electricity needed. Such a system is most likely to be found in a utility-independent residential or commercial installation that has an array capacity of over 1 peak kilowatt.

For all practical purposes, this generator-assisted system represents the highest level of sophistication for stand-alone photovoltaics. However, it can be modified in several ways. For one thing, if a larger-scale stand-alone PV system is desired, the DC load and distribution system may be eliminated in favor of an all-AC system, which is better able to handle a heavy load. Also, other sources of auxiliary power may be integrated into the system. For example, in addition to the PV array and a gen-set, a wind turbine may be added as an auxiliary power source if the wind resource is sufficient at the site.

Utility-Interactive Systems

The traditional utility-interactive (UI) photovoltaic system is one in which electricity is brought in from the grid to supplement the system's output when needed and sold back to the utility when the PV array's output exceeds the on-site load demand. Figure 2-6 depicts a transitional system, one that does not completely stand alone but does not fully interact with the utility grid, either. In this system, power from the grid is substituted for the gen-set as an auxiliary source of electricity.

In this hybrid system, we still have a DC load, served as before by direct array output and the electricity stored in the batteries. And, as with the systems shown in figures 2-4 and 2-5, AC loads are served through an inverter that processes the DC output from the batteries. When both the array output and the level of battery charge are low, power from the grid

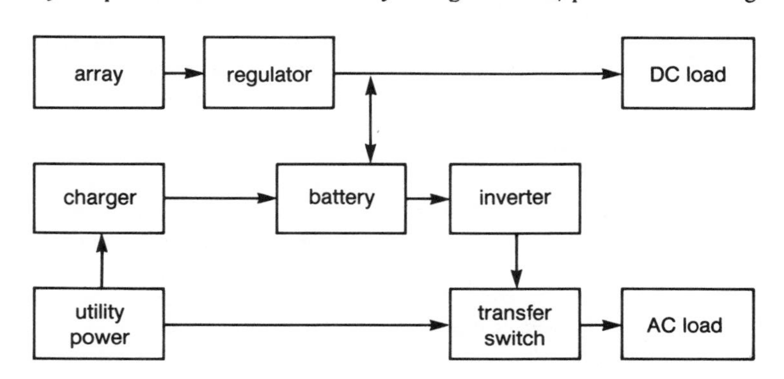

FIGURE 2-6
A schematic of an AC/DC PV system equipped with a voltage regulator, battery storage, and an inverter, which uses the utility grid to supply backup power via a battery charger.

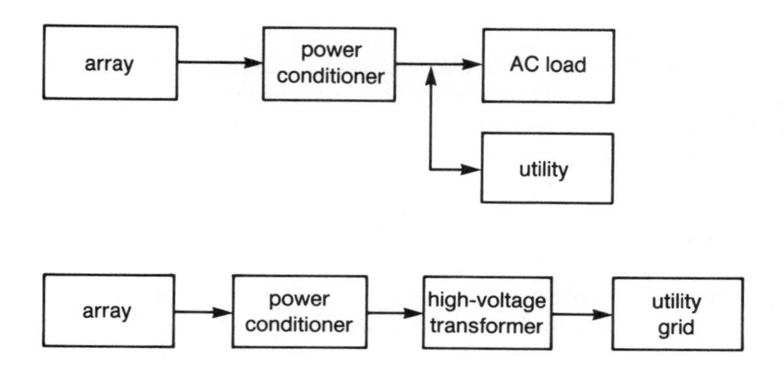

FIGURE 2-7
A schematic of a utility-interactive PV system, which employs a power conditioner designed to make the power produced by the array compatible with that available on the utility grid.

FIGURE 2-8
A schematic of a central-station utility PV system, which employs both a power conditioner and a high-voltage transformer.

comes in to make up for the shortfall. The utility-supplied power flows to the DC side of the system through the battery charger. A simple transfer switch disconnects the inverter, allowing the utility power to handle the AC load directly.

This approach is especially good for someone just beginning to use photovoltaics in an area where utility power is available, because it allows for incremental change. At the outset the PV system can be kept small, with power from the grid playing a substantial role. Over time the size of the array and the output of the PV system can slowly be increased, reducing the amount of auxiliary power needed from the utility. Such a setup is appealing in commercial and residential situations where utility-grid power is reliable and relatively inexpensive but full interaction (that is, sell-back of power) with the utility is not possible because of a poor buy-back rate or other policy problems.

A pure utility-interactive photovoltaic system is a different matter, as the simplified diagram in figure 2-7 shows. Here there is no provision for battery storage of PV output, since none is needed. All loads are AC loads, and consequently the system's entire PV output is converted to alternating current by means of an inverter, or power conditioner, designed especially for use with a UI system. Any shortfall between load demand and array output is automatically met by power from the utility grid. Conversely, all site-generated PV power exceeding the load demand is fed into the grid. The utility must buy this power from the PV system owner at a price equal to its "avoided costs"—that is, the savings it realizes by not having to generate the same amount of power itself at a central station. This is a provision of the Public Utility Regulatory Policies Act of 1978 (PURPA).

The final order of magnitude for earthbound photovoltaic generating systems is represented by the utility-scale, central-station PV array field, such as the California systems described in chapter 1. Here we have photovoltaics put to work to deliver large amounts of power to the central utility grid in much the same way as other large-scale conventional central-station power plants. This scheme calls for all of the DC output from the array fields (which are generally of megawatt range) to be converted to alternating current by one or more power conditioners, and then stepped up to transmission line voltage by high-voltage transformers as shown in

PHOTO 2-1
This multimegawatt, central-station photovoltaic power plant located near San Luis Obispo, California, was constructed by Atlantic Richfield.

figure 2-8. At that point it is fed into the central utility grid and distributed to the utility's customers.

The technological viability of sending power from a central-station photovoltaic array into a utility has already been demonstrated. As the cost of PV systems decreases and the cost of conventional means of power generation increases, the economics of using photovoltaics will become increasingly compelling to utility companies.

Obviously, many variables will determine the speed at which a substantial photovoltaic central-station infrastructure is created in the United States, but it is really not a question of *if,* only of *when.* The future market for PV power plants is sure to be enormous, a fact that should be encouraging to homeowners wishing to install their own PV systems. As use of photovoltaics proliferates in the central-station area, the price of system components will decline, thanks to mass production and increased competition among manufacturers. This will help all other PV market segments, since the same PV cells and modules are used in all applications, whatever their scale.

The ascending order of size and complexity illustrated in these eight system diagrams demonstrates the versatility of photovoltaics. Clearly this is not a monolithic technology. With photovoltaics you can design a system that will deliver as much or as little power as you want, and there are plenty of options from which to choose.

Choosing a Photovoltaic System

Now that we've reviewed the basic photovoltaic system configurations, let's explore the process by which you can choose which option is right for you. As we have seen, the most basic distinction that can be made between systems is whether they stand alone or interact with a utility grid.

A stand-alone photovoltaic system is entirely self-contained, though it may or may not include battery storage or an emergency backup system.

For purposes of definition, we can say that a *stand-alone* photovoltaic system is one that is entirely self-contained, even though it may be supplemented by an emergency backup system supplied with a fuel brought to the site. Such a system can thus be represented as a closed triangle, the components of which are the array, the battery storage bank, and the load.

A *utility-interactive* system, by contrast, is one in which the need for on-site storage is precluded by the system's connection with the local utility grid. The utility not only takes the place of battery storage in such a system, it also serves as a constantly available source of backup power. Systems in which the kilowatt-hour (kWh) meter runs backward (either literally or figuratively) to show sales of excess electricity to the utility are utility-interactive. With this type of arrangement, the photovoltaic user keeps a perpetual two-way account going with the utility company.

A utility-interactive system needs neither battery storage nor emergency backup system, since it is connected with the utility grid.

Each of these two basic options offers some advantages and some disadvantages. Except for pointing out the obvious fact that the unavailability of grid power dictates an SA system, it is impossible to flatly define circumstances or criteria that would mandate the choice of one photovoltaic system configuration over the other. And in cases where both options are possible, the final decision may be philosophical as well as technical. In the most general sense, however, it can be said that the size and type of the load will determine the size and type of the system.

This rule of thumb applies particularly well to the next major decision facing those who opt for stand-alone photovoltaic power: whether to plan your system to deliver alternating current, direct current, or both. Let's look at the choice between the two types of current in relation to SA systems, since the differences between them may influence the larger decision of whether a particular system should be stand-alone or utility-interactive.

Alternating Current or Direct Current?

Despite the universal prevalence of alternating current in the power grids and the domination of the marketplace by appliances built to draw upon alternating current, the fact remains that when it comes to photovoltaics, a choice of current types exists. The reason the DC-powered option is viable for stand-alone PV systems is that when you are not connected to the grid, the normal practice in the outside world does not have to dictate what you do. Less obvious—and more frequently overlooked—is the fact that a substantial number of DC-powered appliances are available and the means exist to convert some of those that use alternating current to DC power.

Many manufacturers are now designing DC-powered appliances and equipment specifically for use with photovoltaics.

Until recently, most available DC-powered equipment was designed for the recreational vehicle (RV) market. However, many companies have now begun to manufacture efficient DC-powered lights, pumps, motors, and appliances for use with photovoltaics and other alternative energy systems. These days, DC-powered equipment compatible with photovoltaics can be purchased through many different outlets, including PV hardware dealers and distributors, marine, camping, automotive, and

RV supply houses. (More detailed information on DC-powered equipment can be found in chapter 6. Some supply sources for DC-powered equipment are given in the list of helpful addresses found in the back of the book.)

However, there are inherent drawbacks to choosing to use only equipment made to operate on direct current. First of all, there are limitations in selection and availability of DC-powered appliances. Though the market is expanding, currently there is little comparison between the amount and diversity of AC-powered equipment and its DC counterparts. The second drawback to an all-DC system has to do with the loads themselves. Large, low-voltage appliances are often less efficient than those built for house current, and low-voltage DC motors that deliver much more than ⅓ or ½ horsepower are impractical. Also, the majority of DC motors, unlike AC induction and synchronous motors, have brushes that wear out and must be periodically replaced, which adds to the maintenance requirements of DC-powered equipment.

The third problem is that direct-current appliances are nearly always low voltage—12 volts is the norm in most systems, as opposed to the 120 volts standard for AC house current. This is fine if the load is going to be kept small, but the high amperage required by loads much greater than 1,000 watts (1 kilowatt) will create problems with wire sizing and voltage drop. To understand why this is so, keep in mind the simple formula *volts times amperes equals watts.* This means that when voltage goes down, amperage must go up to satisfy the power requirements of a given load.

volts (V) × amperes (A) = watts (W)

Wire is rated and sold according to gauges; the smaller the wire, the higher the gauge number. There is a maximum amperage-carrying level for each gauge of wire, above which overheating and significant voltage drops will occur. Thus, in a heavy-demand system the usual strategy is to

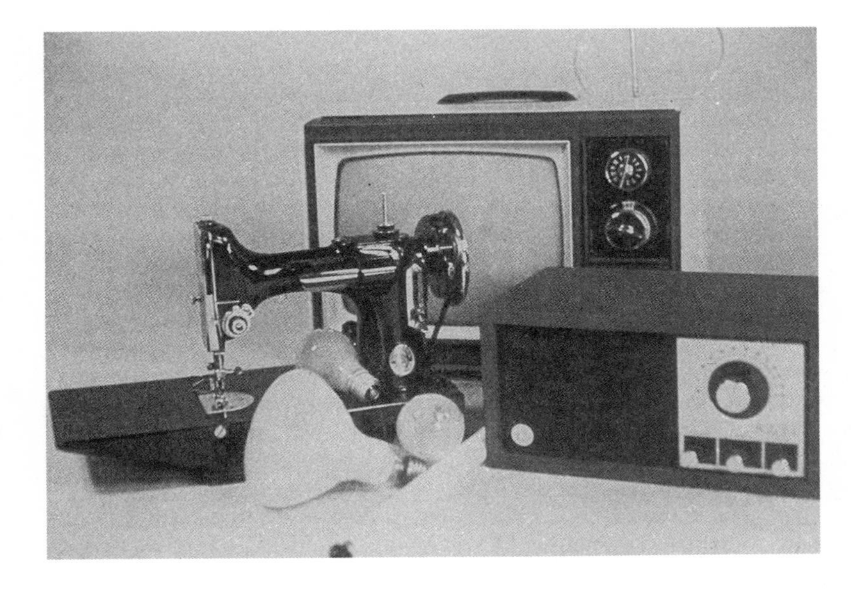

PHOTO 2-2

A wide variety of DC-powered lights, appliances, and tools is available for use with home photovoltaic power systems.

keep the voltage high so the amperage can be kept low. This permits the use of lighter, less expensive wire than would otherwise be necessary.

When low-voltage DC power is to be delivered at the high amperage needed to satisfy the high wattage requirements of "conventional" loads, very heavy wiring must be employed if overheating and voltage drop are to be avoided. Since heavy cable is both more expensive and more difficult to work with, the temptation in a situation like this is to select undersize wire. The fact is that high voltage at comparatively low amperage may be transmitted much more efficiently than low voltage at the high amperage required to deliver the same wattage. (A list of wire sizes, their resistance, and their maximum current-carrying capacity can be found in table 10-1.)

Given all of the drawbacks of DC service, you may well ask, Why consider direct current at all? The answer can be found again in the maxim: The size and type of load will determine the size and type of system. A stand-alone photovoltaic system with DC-only distribution can deliver fine service *if* the system's power output and load requirements stay safely below the point of diminishing returns. Most PV system engineers agree that if the load regularly exceeds 1,000 watts, AC distribution should be considered. On the other hand, if the load can generally be expected to remain under this figure, direct current is a viable choice as long as the desired appliances are available.

Circumstances favorable to the use of direct current are precisely those that prevail for many isolated homesteads and modest vacation retreats now using photovoltaic power. The owners of such residences have formed one of the more important groups of early photovoltaic users. While there is no question that we are entering an era when photovoltaics will become an increasingly realistic and economical alternative for conventionally equipped, full-size homes, the market for small, isolated installations will also continue to grow. If a system is so small that an inverter represents a sizable additional expense, or if the goal is to keep things as simple as possible, a DC-only system may make good sense.

However, heavy loads require higher voltage than light loads, and higher voltage is generally associated with AC rather than DC loads. (Residential AC loads in the United States are typically 110, 115, or 120 volts, in contrast to typical DC loads of 12 volts.) In a moderate- to high-demand SA residential system, the answer may be to opt for alternating current over direct current. But if your load demand is on the borderline, you should consider a combination of AC and DC distribution.

VAC—
volts of alternating current

VDC—
volts of direct current

Making the decision to deliver AC power from an SA photovoltaic system will mean installing an inverter. Inverters can be purchased for a wide range of DC-input voltages, everything from 12 to 120 volts, and they all deliver standard house current: 120 or 120/240 volts of alternating current (VAC). This being the case, you might think the choice of the PV system's battery output voltage becomes arbitrary when it is going to be boosted by an inverter to 120 VAC.

The house wiring used to connect the inverter to the various loads can be sized smaller than the DC wiring because the higher voltage delivers the required power at lower amperage, eliminating the problem of wire losses in distribution. Nonetheless, the choice of the DC voltage at which the photovoltaic array and battery bank will operate is equally as important, or even more so, in a system that includes an inverter as it is in a DC-only system. One obvious consideration is the main line wire connecting the battery bank to the inverter. If an inverter capable of delivering 5 kilowatts (kW) of alternating current at full load is drawing from a 120-volt battery bank, it is doing so at a fraction under 42 amperes. If, however, the same inverter is drawing 5 kilowatts from a 48-volt battery bank, the amperage is 104. And if 24-volt batteries are used, the amperage is nearly 210. In either of the latter two cases, it is easy to see that the line between the batteries and inverter would have to be a very low-gauge, heavy-duty wire.

To keep voltage drop within acceptable limits, current flow from the battery bank to the inverter should be kept at or below a maximum of 100 amps. Thus, for example, if the system is 12 volts, the inverter capacity should not exceed 1.2 kilowatts, while a 24-volt inverter can be sized up to 2.5 kilowatts. There are additional reasons why the photovoltaic system voltage for a stand-alone residence should be sized as high as is practical, especially if an inverter is employed. They all have to do with reducing system losses and improving efficiency.

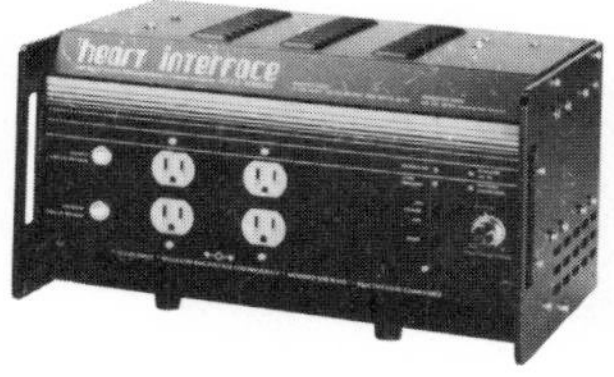

PHOTO 2-3

DC-to-AC inverters can be purchased to satisfy a wide range of power requirements for homes powered by stand-alone PV systems.

The conversion efficiency of an inverter is significantly influenced by the amount of "stepping up" required from the DC input voltage to deliver AC house current. For example, if all other factors are equal, a given inverter will deliver 120/240 VAC at a higher efficiency on a DC input of 48 or 120 volts than on an input of only 12 or 24 volts. Inverters run at their highest efficiency at or near their full rated capacity, so good system design requires careful sizing of the inverter to minimize losses. At the same time, the designer must keep in mind that all inverters have internal losses during operation and while on standby waiting for a load to require power. These considerations lead to the justification for the AC/DC hybrid system.

In most small to moderate-size stand-alone photovoltaic systems that have an inverter, there are many loads on the system that could just as well be served directly by direct current. By configuring the system to deliver *both* alternating current and direct current, we can use each type for the specific loads that require it. This will allow us to keep the inverter's internal losses to a minimum by reducing inverter operating time, and it will probably even allow the installation of a smaller inverter. In addition, a certain measure of redundancy is provided when direct current and alternating current distribution are combined.

If inverters were totally reliable, this redundancy would be undesirable. As improvements are made in inverter technology, this will likely become the case. However, at the present time inverters are the most vulnerable part of a photovoltaic power system, so it is useful not to be totally dependent upon them. A final argument in favor of the hybrid system is the fact that some DC loads, such as certain types of lights and some motors, are actually more efficient than their AC counterparts.

A photovoltaic system can be designed to provide DC power directly to loads able to run on it, while converting some power to alternating current for loads that specifically require it.

Examples of loads that run on direct current are fluorescent and incandescent lighting, well pumps, refrigerators, evaporative coolers, and many other appliances. (These will be discussed in detail in chapter 6.) In addition, certain appliances and equipment can run on either alternating current or direct current at 120 volts. These include most hand power tools, many small appliances, and some vacuum cleaners—basically, any pieces of equipment that use universal motors. Tools and appliances with universal motors can provide added flexibility to an SA photovoltaic residential system if the loads on the system are large enough to justify a DC voltage of 120 volts.

At present, then, the preferred system design for the majority of moderate-demand, stand-alone photovoltaic-powered homes with a connected load greater than 1,000 watts will probably include both AC and DC service. However, as the state of the art in SA inverters continues to improve, I expect more and more PV homes to opt for the AC-only approach, with DC distribution continuing to be used only in small, remote cabin power systems.

Utility-Interactive or Stand-Alone?

The tougher question facing the would-be photovoltaic homeowner is whether to opt for stand-alone or utility-interactive PV power. Unlike the alternating current versus direct current question, this one cannot be settled simply by following a size-related rule of thumb. A variety of issues are involved.

Stand-alone photovoltaic systems are often preferable in rugged or remote locations where a grid connection is extremely expensive.

Distance from the grid is one important factor to consider. In some places—remote islands, for example—grid connections are simply out of the question. In other locations, utility companies will contract to provide a hookup upon payment of an initial fee, which may be coupled to a schedule of minimum monthly electric bills over a set period of time. If the line is not too far away, the cost may not seem excessive relative to the installed price of a photovoltaic system. But if your homesite is located more than a half mile from the nearest utility connection or is separated from that connection by unusually rough terrain, a power company hookup will probably prove very expensive. When comparing the cost of a new grid connection with the cost of installing a PV system, also remember that the utility bills keep coming every month, forever.

Even when the choice of an SA system is not ordained by the impossibility or expense of grid connection, many practical considerations argue in its favor. Perhaps the most important is that a stand-alone system is *independent* of the utility and will provide you with power even during times of grid failure. This is not the case with UI systems, the vast majority of which have no on-site electrical storage and depend upon line-tied inverters, which cannot operate without the presence of the utility power.

But it would not be fair to stress only the advantages of an SA system, since there are several practical disadvantages—or at least inconveniences—worthy of note. Battery storage is a necessity, which adds to the system's expense, space demands, and maintenance requirements. In addition, auxiliary generating equipment is often needed for occasional replenishing of battery charges during heavy use or prolonged periods of poor insolation. Furthermore, each time additional photovoltaic modules are added to the array, storage capacity must be increased proportionally. And, periodically, more power may be produced than the system can immediately use or store in its batteries. Since it cannot be sold to the utility, this power will be wasted.

Utility-interactive systems also offer both positive and negative benefits. Let's look first at some of the advantages that they offer in comparison to stand-alone systems.

- Grid connections eliminate the need for battery storage, unless some emergency reserve is desired (more about this later).
- Excess power produced by the system can be sold to the utility.
- The grid itself constitutes the backup when the photovoltaic system's output is not sufficient to satisfy the load, or in the unlikely event the array fails to operate.
- The number of modules in the array—and consequently its electrical output—may be increased without increasing storage capacity (although this may necessitate replacement of the inverter with a more powerful model).
- The PV system need not be sized to accommodate the load's peak and occasional high surge requirements.
- The system can be sized to handle any desired portion of the total load requirements, from a small percentage to over 100 percent.

Despite their many advantages, UI systems are not without drawbacks. Here are two to consider:

- The systems configurations currently available cannot operate when utility power is down.
- Interactive systems require that electrical current variables such as frequency, power factor, harmonics, and wave shape be much more carefully regulated than is necessary for SA systems, in order to assure compatibility with utility-supplied line current. This requires high-quality power-conditioning equipment, which is more expensive than the equipment needed for stand-alone systems.

After weighing the pros and cons of both options, we may reasonably conclude that at sites where power lines are easily accessible, a combination system can provide us with the best of both worlds. Ideally, such a system would be equipped with a self-commutated inverter (that is, an inverter able to reach and maintain the appropriate AC frequency without stimulation from an outside source) capable of both UI and SA operation. In fact, one inverter manufacturer has developed a prototype of this type of unit and is considering its introduction on the market.

Another component that would be included in an ideal hybrid installation is a small standby battery bank. When the photovoltaic system is interacting with the utility, the batteries would be "floated" at constant full

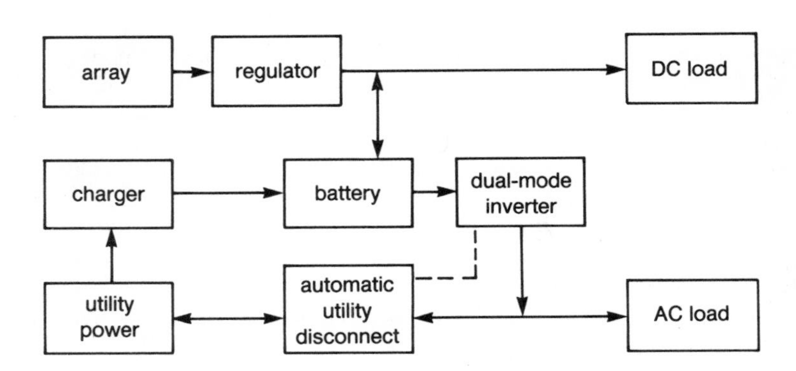

FIGURE 2-9
A schematic of a dual-mode system, which employs a voltage regulator, battery storage, and a dual-mode inverter capable of either utility-interactive or stand-alone operation and features an automatic disconnect to prevent PV system power from flowing out onto the utility line during a line failure.

charge. Then, during a utility blackout or at any time when the homeowner wished to isolate the PV system from the grid, the dual-mode inverter would convert the DC battery power to alternating current for the house loads. (A diagram of this type of system is provided in figure 2-9.)

But would such a setup actually be as ideal as it sounds? Keep in mind that any system that takes advantage of both options also incurs the combined expense. The battery bank in our hypothetical system would have to be quite large to provide real insurance against a prolonged grid outage. Most power failures occur either at night or during bad weather, when photovoltaic array output is insufficient either to serve loads directly or to provide much of a battery charge. Thus, if the battery bank were small, the house would quickly run short of electricity unless the normal load demand was reduced. But if the battery bank were made large enough to carry the entire household demand through a prolonged emergency, what advantage would this system offer over a simple SA system?

The truth is that the benefits derived from combining the SA and UI systems into one are more psychological than economic. They will be most appreciated by the homeowner who wants to enjoy the benefits of utility-supplied power along with the security of having a backup power source to call upon in times of need. Such a person may live in an area plagued by frequent power outages or may simply desire the ability to control his own energy destiny and provide for his family during a greater emergency.

In the final analysis, the decision between a stand-alone and a utility-interactive photovoltaic system involves more than a straightforward, rational exercise in weighing pros and cons. Personal ideals and preferences must be taken into account. These systems are not merely two ways to implement a new and exciting technology. They also represent different expressions of the relationship between the individual and the community. One approach expresses a commitment to self-reliance and independence, the other a willingness to be linked, for better or worse, with a larger community of energy consumers.

Those who make a commitment to SA photovoltaics argue that one of its principal attractions is its promise of truly energy-independent living. While it can be argued that dependence on any sort of fossil-fuel–fired

PHOTO 2-4

This remote vacation cabin receives its power from a small stand-alone photovoltaic system.

emergency backup generator qualifies this promise, the fact remains that the stand-alone advocates are essentially correct. A PV array in tandem with a battery storage bank, with or without an inverter, constitutes the best method available for achieving domestic energy autonomy.

Of course, a person could choose to live without electricity in a log cabin built amidst ten acres of hardwood and achieve energy autonomy in that way. But by adding an SA photovoltaic system to that cabin, energy independence is possible without a total sacrifice of late twentieth-century comforts. Why, the stand-alone enthusiasts ask, should we take such a splendidly self-contained operation and link it to the very power network it enabled us to become free of? Worse yet, why hand control of our current inverters over to that grid, making site-produced PV power inaccessible in the event of a utility grid failure?

Instead of a community of autonomous mini-utilities, the proponents of UI photovoltaic power envision a network of power-producing systems interactive not only with the commercial grid but also with each other. The first community of this type was recently completed in Gardner, Massachusetts. Dispersed grid-connected PV systems totaling 100 kilowatts of peak capacity installed on existing residences, as well as on commercial and municipal buildings, all deposit PV-generated power into one distribution feeder on Massachusetts Electric Company's grid.

A photovoltaic array combined with battery storage offers the individual homeowner an opportunity to achieve domestic energy autonomy.

This experiment in community photovoltaics, funded by the New England Electric System, has proved conclusively that a happy marriage can be made between many distributed PV systems and the utility grid. It has brought the power industry an important step closer to the day when numerous small PV systems working together will furnish a substantial percentage of the power available on a grid. That day is no longer far in the future.

The person who opts for a UI photovoltaic system, then, is likely to see his or her contribution toward energy independence in societal

PHOTO 2-5

This photovoltaic array atop the city hall in Gardner, Massachusetts is part of a 100-kilowatt demonstration program funded by the New England Electric System.

rather than individual terms—although it is certainly true that every SA system installed helps alleviate the energy problems of the nation and the world. The purchaser of a grid-tied array may also be motivated by the desire to spend a minimal amount of time on system overseeing and maintenance. Such maintenance is not particularly time-consuming with a stand-alone system, but it is virtually eliminated with a utility-interactive one. Still, all philosophical considerations aside, the most important factor behind most people's choice of systems is economics.

Photovoltaic Economics

Given the considerable variations in insolation levels that exist across the United States, some readers may assume that there are some places where photovoltaics is a viable proposition and other places where it is not. However, the situation is not that simple. Several variables besides insolation levels help determine whether or not a decision for photovoltaics makes sense in any particular area. Some of these variables may, in fact, lead to a positive decision on photovoltaic installation in a location that, at first glance, does not appear to have adequate solar resources.

The most important of these nonsolar variables is the regional cost of electrical power generated by conventional means and available through established utility grids. This is represented in utility charges per kilowatt-hour, a figure that varies throughout the country much more than the quality and volume of solar radiation, as is evident from the data found in table

Table 2-1. Residential Electric Rates for Selected Areas of the United States

Area	Cost/kWh (¢)	Utility Company
Albuquerque, N.M.	10.0	Public Service Company of New Mexico
Atlanta, Ga.	6.5	Georgia Power Company
Atlantic City, N.J.	11.0	Atlantic City Electric Company
Boston, Mass.	12.7	Boston Edison Company
Chicago Heights, Ill.	13.0	Commonwealth Edison Company
Denver, Colo.	7.4	Public Service Company of Colorado
Hilo, Hawaii	14.3	Hawaii Electric Light Company
Houston, Tex.	8.1	Houston Lighting and Power Company
Miami, Fla.	9.0	Florida Power and Light Company
New Haven, Conn.	12.5	United Illuminating Company
New York, N.Y.	15.7	Consolidated Edison Company of New York
Philadelphia, Pa.	10.8	Philadelphia Electric Company
San Diego, Calif.	13.0	San Diego Gas and Electric Company
Scottsdale, Ariz.	9.7	Arizona Public Service Company
Wilmington, Del.	11.0	Delaware Power and Light Company

SOURCE: Data supplied by U.S. Department of Energy.
NOTE: Data based on January 1985 rates for average monthly consumption of 500 kilowatt-hours.

2-1. In the Pacific Northwest, abundant hydroelectric power has long meant low utility rates—as little as 3 to 4 cents per kilowatt-hour as of this writing. In other parts of the country, rates of 14 cents per kilowatt-hour and higher are far more common. And in remote areas, the cost of grid-supplied electricity will often exceed 20 or even 30 cents per kilowatt-hour. For example, on Block Island, a summer resort in Long Island Sound with a modest year-round population, customers of the local utility pay 35 cents per kilowatt-hour to keep their lights burning and appliances working.

The facts about the economics of utility-interactive photovoltaics should be squared with present realities. Currently, homeowners who are already connected to a power grid cannot realistically think of photovoltaics as a money-saving alternative, regardless of the cost of utility-supplied electricity in their area. But this situation will surely change as photovoltaic costs decrease and utility rates escalate.

It has been estimated that, without factoring in potential tax credits, utility-interactive photovoltaics will become "economical" when installed costs drop to $4 per peak watt in Southern California or $2 per peak watt in Virginia. This is based on an estimate of 2,160 kilowatt-hours of annual PV output and utility rates of 15 cents per kilowatt-hour in the first instance and 1,440 kilowatt-hours of PV output and rates of 10 cents per kilowatt-hour in the second.

There is a simple way to calculate how close photovoltaics is to the economic break-even point in your area. Just take the projected annual output, in kilowatt-hours, of the system you are contemplating, multiply it by a design life of 25 years and divide by the overall system cost, making the necessary accommodations for tax credits and loan interest. Then compare this figure with the current utility rates that apply in your area.

Often there is a seesaw relationship between one variable and another. Residents of New England, for instance, are the beneficiaries of only about half as much insolation as prevails in the sunniest parts of the country. They thus face a proportional increase in the cost of installing adequately sized photovoltaic systems. But the potential savings in utility bills those systems make possible for New Englanders may well be greater than those that could be realized by homeowners in many places blessed with higher average insolation levels. Photovoltaic system costs for the latter group may be lower than those faced by New England residents, but *competing electric rates are also lower,* reducing the incentive to turn to alternative sources of power.

Only in areas where low utility rates are coupled with poor insolation—the Pacific Northwest readily comes to mind—does solar electricity clearly appear to be an economically unattractive proposition. However, as the production of photovoltaic hardware expands, the cost per peak watt of generating capacity should decline. Eventually, we may see PV arrays adorning the roofs of houses even in the land of the Douglas fir. Elsewhere, in more favorable climes, homeowners, builders, and utilities with an eye toward economy as well as innovation will begin to take a serious look at the new technology once the costs of installing PV systems are projected into realistic payback periods.

Photovoltaics in the Sunbelt

Let's look at the economics of a hypothetical 2-kilowatt, UI photovoltaic system. Consider a $20,000 system, constructed in California in 1985, on which a total of $6,500 in tax credits was applied. Using a simple cost equation and figuring in maintenance and interest costs against an annual system output of 4,000 kilowatt-hours and a 25-year system life, this system would deliver electricity at a cost of about 19 cents per kilowatt-hour. This figure is less than a factor of two higher than the average American homeowner's electric bill. Without any tax credits, this figure would increase to approximately 26 cents per kilowatt-hour, or less than a factor of three higher than the current cost of utility-supplied kilowatt-hours in many parts of the country.

California is one of the sunniest parts of the country, so it would appear to be well suited to the use of photovoltaic electricity. But how do the figures look in other parts of the country? They begin to look somewhat less favorable as we start encountering the "moving target" factors that influence our calculations of costs per kilowatt-hour. Foremost among these frequently changing factors are insolation, which determines the number of kilowatt-hours per year a system can produce, and tax credits, which may be retained or discontinued according to the political winds that prevail in any given period of time.

Chief among the "moving target" factors that influence the cost of solar electricity are the amount of insolation and the availability of tax credits in a given area.

If the costs per kilowatt-hour of solar electricity appear to vary considerably in different parts of the country, remember that so do the utility electric rates. And once again, we must remember that the cost of utility power is going up everywhere, while the cost of photovoltaic systems is falling. Also, as we shall see in the following pages, the comparative relationship that exists between utility rates and the cost of PV systems is not the only factor to consider when determining whether photovoltaics is economically feasible.

Photovoltaics in Remote Sites

Currently, the most economically attractive applications for photovoltaics are those situations in which the chosen site is not serviced by grid power. In such situations, the question is one of PV installation costs versus dependence on a diesel generator or (if possible) getting the grid extended. How do the numbers look in this type of situation?

First, let's consider the diesel alternative, which is currently the most popular option in places where electricity must be produced on the site. A diesel generator is an expensive piece of equipment with a predictable life span; its maintenance and replacement costs must be factored into any calculation of its long-term expense, as must the price of the fuel on which it runs. The life span of an 8-kilowatt, 1,800-revolutions-per-minute (RPM) diesel generator operated at one-half load is 7,500 hours; at full load it is 5,000 hours. With installation, an 8-kilowatt generator is likely to cost between $5,000 and $6,000. (Generators operating at 3,600 RPM are less expensive, but they wear out faster.)

Now we arrive at an interesting point, which has a great deal to do with the life span and expense of a diesel generator. It involves the amount of time the machine is operated and the load that it is serving while in operation. Any power source must be designed to supply electricity to the full range of loads on-site. But how often are all of these loads in operation at once? Much of the time, a house load consists of a handful of lights and small appliances operating at, for example, 800 watts or less. That's 10 percent of the rated capacity of an 8-kilowatt generator. Using typical figures for fuel economy and the relationship of workload to engine life, we find that, in terms of cost per delivered kilowatt-hour, it costs approximately six times as much to run a generator loaded at 10 percent of capacity as it does to operate a fully loaded machine (see figure 2-10).

The reason for this is that fuel consumption is not in arithmetical lockstep with load percentage. Neither are maintenance and repair. The evidence is clear: running a generator for long periods at loads well below

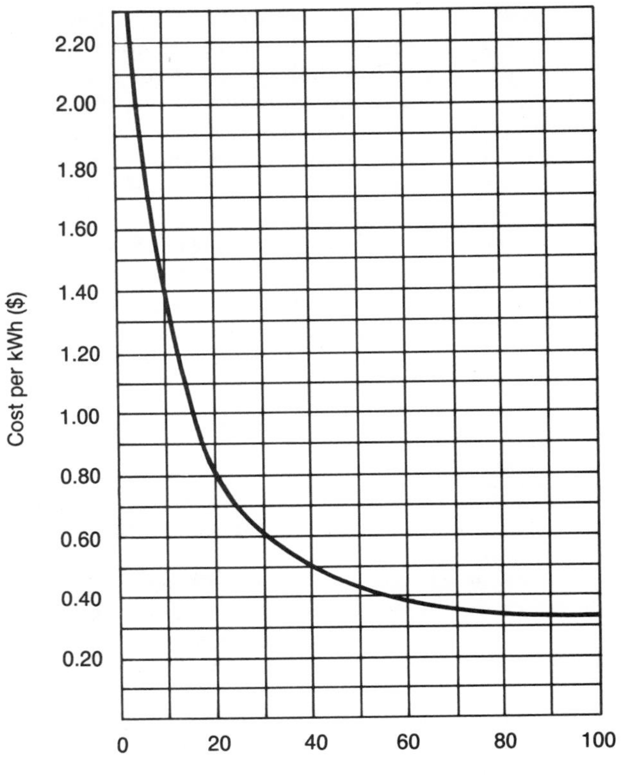

FIGURE 2-10

A graph showing the cost per kilowatt-hour of electricity delivered by an 8-kilowatt backup generator in relation to the percent of rated capacity at which the generator is operated.

It is much more efficient and economical to run a generator at full capacity for short periods of time than at low capacity for long periods of time.

its rated capacity is a very expensive proposition. The figures tell the story all too well. If an 8-kilowatt diesel generator is providing an average of 1,000 kilowatt-hours per month, its average continuous load (24 hours a day for 30 days) is about 1.4 kilowatts (1,000 kWh ÷ 720 hr). That is about 17.5 percent of its capacity. Given standard life span estimates and repair frequency on a $6,000 machine, and a fuel cost of $1.50 per gallon, we arrive at a cost of over $1.00 per kilowatt-hour. The average electric bill at that rate? A thousand dollars a month.

The only way around this dilemma, short of drastically reducing loads and/or imposing a stringent program of voluntary load management that clusters loads at a specific time of day, is to treat the generator as an *auxiliary* source of power. This way the generator has to be used only occasionally. But when it is used, it is run at full capacity for the relatively short period required to restore the charge of a bank of batteries that are primarily supplied by another source such as a photovoltaic array. This combination of short-duration, full-capacity generator use means a longer life and lower kilowatt-hour costs from the machine, since you won't be running it just to keep a few lights and the television on.

It is not my purpose here to denigrate independent power systems that run on fossil fuels; clearly, they can do some things that photovoltaics cannot, and they continue to have their place in specialized applications. But I do suggest that anyone considering such an application take a hard look at the economics of unassisted diesel (or propane) generators and consider the advantages of the combined photovoltaic/generator/battery option. A well-designed system incorporating a PV array, battery storage, and backup generator will optimize the performance of each component.

Of course, dependence on a stand-alone power system, whether photovoltaic or a conventional engine generator, is not the only solution available to people considering living in some out-of-the-way place. In many cases, an extension of the power grid may be an economically viable alternative when compared with the cost of installing a photovoltaic system.

If the distances to be covered are not too great, a utility company will usually agree to extend power lines to the prospective customer's point of use. This will not be done, however, solely out of the firm's anticipation of a long and happy relationship with the new customer. Charges for extending the line will be levied, and these will vary enormously in accordance with the distance to be covered *and* the type of terrain that the lines must cross. Some installations will involve nothing more than the clearing of a few branches over a few hundred yards of relatively level ground; others will require that the installers contend with far more difficult conditions.

At the extreme are situations in which local ordinances require the burying of electrical cables; if a substantial amount of rock has to be disturbed (for instance, if ledge lies too close to the surface), the cost will be very high, even for short distances. With so many determining factors at work, engineering dollar-and-cent estimates for this procedure becomes a tricky business; however, it is fair to say that there are always at least four figures to the left of the decimal, and five are not uncommon even over a short distance.

Of course, if several householders are to live at the end of the same projected line, costs can be shared, but a custom grid hookup is seldom cheap and is usually accompanied by a requirement that the customer purchase a minimum amount of electricity each month for a long time to come. Often, too, the new customer will be responsible for maintaining the grid extension to his site, especially if the service is underground. When both the cost of a grid connection and the projected monthly utility bill over the life of a house are totaled, an SA photovoltaic installation often becomes very attractive.

Photovoltaic system engineers Anitra Sorensen and Robert Wills of Skyline Engineering in Ely, Vermont, have performed detailed cost analyses on the competitiveness of stand-alone photovoltaic systems versus utility-line extensions and generator-only remote power systems. In their report "The Economics of Remote Photovoltaics/Generator Systems" (International Solar Energy Society, 1985), they concluded that PV systems were cost-competitive at 1985 prices at sites no more than 2,500 to 3,000

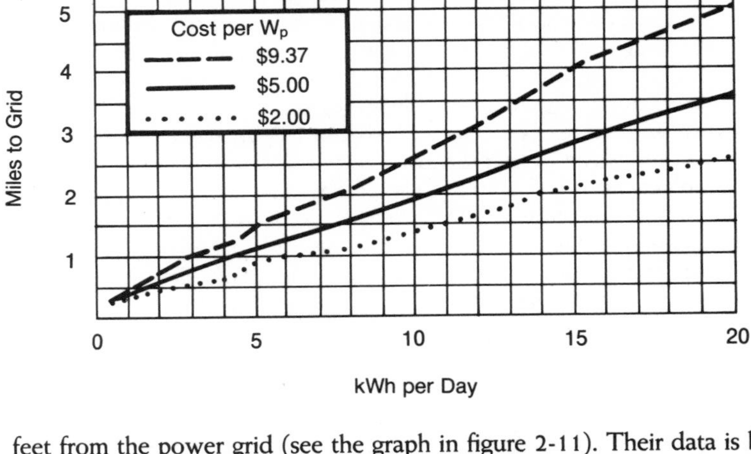

FIGURE 2-11

A graph of cost-effectiveness for stand-alone PV systems (at three different peak-watt cost levels) based on kilowatt-hours per day required versus distance from the existing power grid.

feet from the power grid (see the graph in figure 2-11). Their data is based on small to medium-size PV systems, providing less than 5 kilowatt-hours per day, that they have installed in northern New England, an area not known for an abundance of insolation. In areas with more sunshine, the numbers will look even better.

There is another, more hidden advantage to choosing photovoltaics for a remote installation that may not immediately occur to many readers, but which nevertheless may be translated into real overall cost savings. It has to do with the value of developable land served by, or reasonably accessible to, an established utility grid, versus that which has little or no prospect for grid hookup. Take a look at realtors' listings for country properties and you will see that acreage on the grid is significantly more expensive than remote, unconnected sites. The same price differences are associated with the availability of or lack of water. If there is insufficient water, you are simply out of luck, but you can produce electricity on-site, *while taking advantage of lower land prices.*

Photovoltaics allows the prospective homeowner the freedom to take advantage of inexpensive land located far from the utility grid.

Thus, with photovoltaics as an option, the homeowner has more freedom in choosing where he wants to live. Often a site that is much more desirable than land served by the grid can be found at what would be considered bargain-basement prices. Money saved on the purchase of land can be applied to the installation of a photovoltaic system and may justifiably be deducted from the cost of that system when the figures for projected payback time are laid out and analyzed. This concept has even more appeal when more than one house is being considered. There is the possibility of sharing the costs of the power system, especially when careful load management is part of the system design. In years to come, perceptive builders and developers will come to understand the value of photovoltaics as a development tool, and entire energy-independent villages will be built that have no connection to the utility grid.

We have seen that photovoltaics is an extremely flexible technology that provides a multitude of options for the supply and use of electricity. For residential applications, you can choose a PV system to deliver as

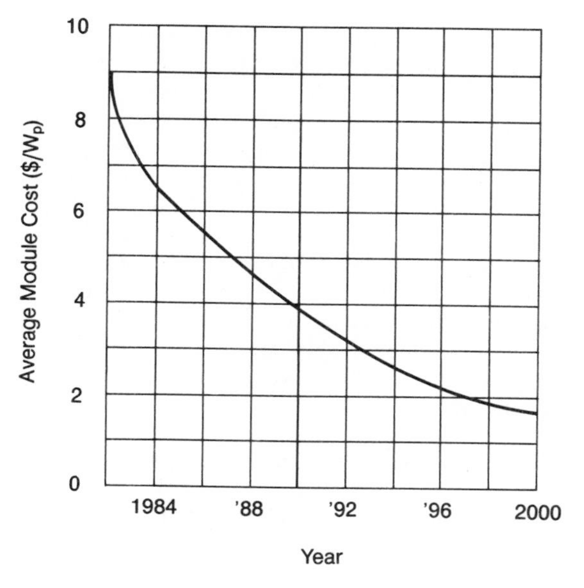

FIGURE 2-12

Over the past ten years the cost of photovoltaics has dropped dramatically. Experts predict module costs under $2 per peak watt by the year 2000.

much or as little power as you want, ranging from a small DC system to power the lights in your hunting cabin to a fully automatic AC/DC system that provides you with all of the amenities enjoyed with utility-grid power. Not only does this technology offer you flexibility in terms of system size and type of current, it enables you to control the degree to which you become dependent upon the utility grid, if at all. With a stand-alone PV system, you can choose to live anywhere you like regardless of the availability of a utility-grid connection. On the other hand, a utility-interactive system allows you to produce your own electricity on-site and sell your surplus back to the utility company.

Some of the applications discussed in this chapter are cost-effective today and some will not be for a number of years. How many years? Photovoltaics expert Paul Maycock predicts that the price of PV modules will drop by nearly 50 percent over the next five years and then continue downward through the 1990s as sales volume increases, production processes are improved, and new technologies are introduced into the marketplace. (See the graph in figure 2-12.) Of course, each application of photovoltaics has its own individual set of economic criteria and its own point of cost-effectiveness. However, the market for photovoltaics in general will grow geometrically as application after application achieves widespread acceptance.

Currently, the installed cost of residential photovoltaic systems ranges from $9 to $12 per peak watt for UI systems, from $12 to $16 per peak watt for utility-backed systems, and from $14 to $22 per peak watt for SA systems. Prices will, of course, vary with system size and degree of sophistication. The higher-priced stand-alone systems include a fully automatic system with extended storage and an industrial-grade backup generator. A small SA system capable of powering lights, entertainment equipment, a few small appliances, and perhaps a shallow-well pump will cost between $6,000 and $10,000. A 1-kilowatt stand-alone system costing approximately $18,000 will deliver electricity at between 45 and 60 cents per kilowatt-hour using a "simple" payback, depending on the site, the choice of components, and the estimated value of future electricity generation.

Although the schizophrenic nature of government support has affected the pace of development of the photovoltaic industry within the United States, it will continue to expand rapidly in the future, because its growth does not depend on events within this country alone. Already, many other countries have embarked upon their own PV development programs: China, Canada, England, West Germany, France, Saudi Arabia, and Japan, among others. Photovoltaics may not yet be a household word and PV products may not yet be available at your local hardware store; nevertheless, this industry is growing rapidly all over the world and is on the verge of making a substantial penetration of the world energy market.

This is especially true in Japan, where a great deal of attention is being given to photovoltaics. Thus, if the American PV industry loses its lead and momentum due to the lack of adequate government support, the United States may end up exchanging its dependence on imported Middle East oil for a dependence on imported Far Eastern photovoltaics.

Given these trends, and the problematic future of so many conventional energy sources, we have to wonder how much longer the word *alternative* will remain part of the photovoltaic vocabulary. There is no need to speculate about when the solar electric future will be upon us. It has already arrived. With photovoltaics, the future is now.

CHAPTER THREE

PHOTOVOLTAIC MODULES AND ARRAYS

Creating a complete photovoltaic (PV) system involves a series of design steps. One of the first steps is to determine the output of individual photovoltaic modules and figure out how to assemble them into a functional array. That is the focus of this chapter. Subsequent chapters will describe the additional components needed to complete a system and will explain how systems should be sized, installed, operated, and maintained.

Combining Cells into Modules

In chapters 1 and 2, you were introduced to some basic electrical terms—amperage, voltage, and wattage—that are important for understanding how solar cells and modules work. Voltage, you will recall, is the measurement of electromotive force, or electrical "pressure." In each individual solar cell, this pressure remains nearly constant regardless of the cell's size. The way to increase voltage, then, is not to increase the size of each cell but to link several cells together. A specific number of cells are connected in series—that is, positive to negative or back contact to front contact.

Amperage, or current flow, on the other hand, is a function of cell area and the amount of light to which the cell is exposed. But amperage is also best increased by linking groups of cells rather than by producing impractically large and expensive cells. In this case, individual cells are joined in parallel—that is, negative to negative and positive to positive, with contacts joined front to front and back to back.

When a group of photovoltaic cells that are joined in a series or parallel configuration, or both, is combined into one sealed unit, what results is the PV module, the basic power-producing unit in all PV energy systems. All of the PV modules that deliver electrical energy to the same load are collectively referred to as the PV array. A group of modules

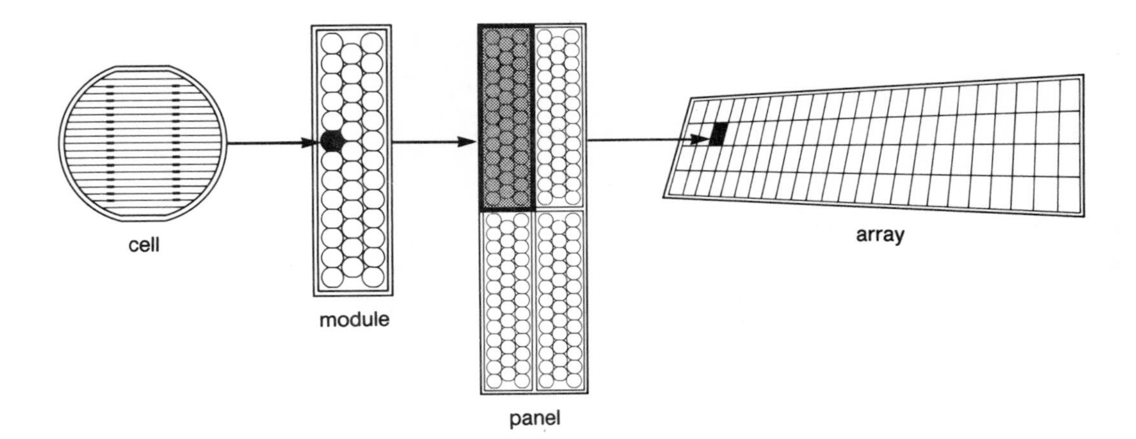

FIGURE 3-1
A diagram showing the relationship among a photovoltaic cell, module, panel, and array.

within the array connected and mounted together to form a single structural component is sometimes called a *panel*. The PV modules wired together to form one series circuit within a PV array are often called a *series string* or *array string*, and are sometimes referred to as a *source circuit*. The relationship between these different parts of a PV array is shown in figure 3-1.

Photovoltaic arrays can be assembled across a very broad range of power requirements to closely match the needs of any specific application. Representing the very small scale are PV arrays that deliver milliwatts of power to watches and other consumer electronic items. On the other end of the power spectrum are arrays that pump megawatts of power into the central distribution grids of the world's largest electric utilities. This nearly infinite flexibility is one inherent attribute of photovoltaics that distinguishes it from all other sources of electricity and makes it so very appealing.

Photovoltaic Modules and Module Ratings

The amount of power produced by a photovoltaic module or array and delivered to the load is dependent upon many factors. The two most important ones are the intensity of sunlight striking the cells and the operating temperature of the cells. A pair of additional factors are the matching of the output of the array with the electrical load and the internal electrical matching losses between individual cells in a module and individual modules in an array.

The fact that the output of a photovoltaic module or array is directly proportional to the amount of sunlight it receives is fairly obvious. The second important factor—namely, the effect of temperature on the performance of a PV system—is not so obvious. Unlike solar thermal systems, PV devices are most productive when they are cold; they produce less and less electricity as their operating temperature increases. In addition, the relative thickness and composition of the earth's atmosphere influences the output of a PV cell or module by selectively filtering out certain portions of the solar spectrum. Because of these and other variables, a common set of criteria is needed for describing PV hardware.

Fortunately, the U.S. Department of Energy (commonly known as DOE) has created an industry standard for testing and rating photovoltaic hardware based on performance specifications established for DOE-sponsored PV projects. One of the most basic and important of these rating terms applies to the measurement of the power or wattage output of PV cells, modules, and arrays. It is the *peak watt,* sometimes referred to as *watt(s) peak* and abbreviated W_p. The peak wattage of a photovoltaic device is defined as the amount of power it can be expected to produce at noon on an (ideal) cold, clear day under full, bright sun. When the output of a large PV array is the subject of discussion, the term *peak kilowatt(s),* or *kilowatt(s) peak* (kW_p), is frequently used.

The DOE, the U.S. Department of Energy, has created an industry standard for testing and rating photovoltaic hardware.

In order for every manufacturer's peak-watt rating to mean the same thing, the industry, under the guidance and direction of the Department of Energy's photovoltaic program, agreed upon a set of *standard test conditions* (STC) under which the peak power rating of their modules would be determined. This standard was established for the DOE by the Jet Propulsion Laboratory (JPL) at the California Institute of Technology and is defined in terms of the operating cell temperature and the amount and type of solar energy available for conversion.

***Peak watt (Wp)*—the amount of power a photovoltaic device can be expected to produce at noon on a cold, clear day under full, bright sun**

The specified operating cell temperature under standard test conditions is 25°C (77°F). The measure for solar insolation is expressed in watts per square meter (W/m²) of module area with a value for *air mass* (AM). The idea behind the air mass figure is simple: since the earth's atmosphere interferes to a varying degree with the transmission of the solar spectrum, some means of expressing this interference is necessary. In space, where there is no atmosphere, the air mass value is set at 0. On earth, an ideal air mass value would be called 1, but the ideal situation is seldom found. It is usually mitigated by the presence of haze, dust, water vapor, pollution, or any combination of such agents. The JPL-mandated STC for photovoltaic testing, then, includes an air mass value of 1.5 and an irradiance of 1,000 watts of solar energy falling on each square meter of module surface. In shorthand, the formula is 1,000 W/m² @ AM 1.5.

***Standard test conditions (STC)*—a set of guidelines established for DOE by the Jet Propulsion Laboratory for rating the peak watts of PV modules; they include cell temperature of 25°C (77°F), insolation of 1,000 watts per square meter, and air mass of 1.5**

This peak-watt rating is very good for comparative purposes but is unfortunately of little direct value in PV system design because the parameters under which it is taken represent an unrealistically high electrical output. This is because under typical field conditions, the sun seldom shines on a module at 1,000 watts per square meter, and cell operating temperature is seldom as low as 25°C. Thus, the output of your PV modules will probably be lower.

To account for the effects of temperature and create a module rating that more closely represents the electrical output under actual field conditions, the JPL established another set of criteria called *standard operat-*

ing conditions (SOC) and *nominal operating cell temperature* (NOCT), along with several other systematic measurements of insolation and environment.

Standard operating conditions (SOC)—a set of ambient microclimatological circumstances that include insolation of 800 watts per square meter, air temperature of 20°C (68°F), wind velocity of 1 meter per second, the module oriented toward solar noon, and open-circuit voltage

Nominal operating cell temperature (NOCT)—the temperature reached by a cell while it is functioning under standard operating conditions

Nominal operating cell temperatures vary for different cells and modules and for the same module in different mounting configurations. According to JPL specifications, the NOCT will be defined as the temperature reached by a cell while it is functioning under standard operating conditions. SOC, in turn, consist of the following set of ambient microclimatological circumstances: insolation of 800 watts per square meter; air temperature of 20°C (68°F); a wind velocity of 1 meter per second; the module mounted in its final position and oriented toward solar noon; and open-circuit voltage—that is, no current flowing. This set of conditions much more closely approximates what the photovoltaic module will experience under actual field conditions than do the standard test conditions described earlier.

The impact of cell operating temperature on cell output is an important variable. All photovoltaic modules experience a drop in operating voltage as temperature rises. This results in roughly a 0.4 percent reduction in power output for each degree Celsius of increase in cell temperature. The NOCT of a module in the field will vary from 20 to 40 degrees Celsius above the ambient temperature, depending on the design and packaging of the module, the manner in which the module is mounted, and the amount of insolation it is receiving. In virtually every installation, the NOCT will be substantially higher than the 25°C specified for STC test purposes, resulting in a module power output lower than the peak-watt rating at STC. For example, a module that produces 40 peak watts under STC will deliver only 36.8 peak watts at a more normal operating cell temperature of 45°C (113°F) under insolation of 1,000 watts per square meter. When the insolation level is reduced to the 800 watts per square meter specified in SOC, which more closely represents actual average field conditions, the module's output will drop to about 28 peak watts. The effect of temperature on PV output is shown graphically in figure 3-2.

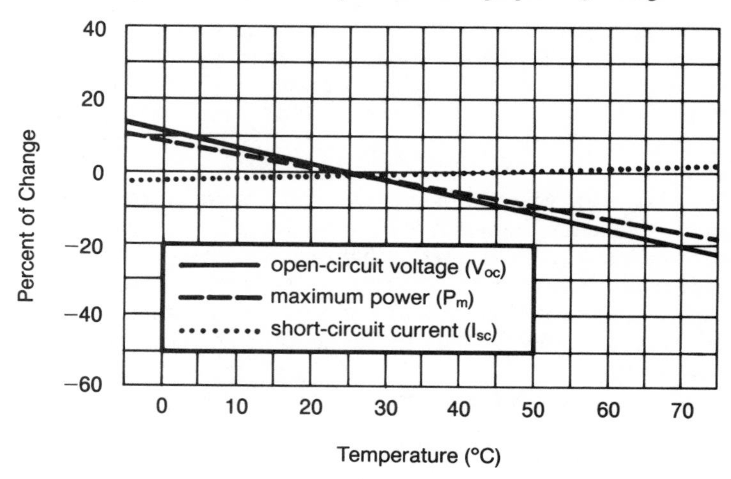

FIGURE 3-2
A graph showing the effect of temperature on photovoltaic module performance.

In practice, the impact of temperature on photovoltaic array output is most significant on large, high-voltage systems. On smaller, battery-charging systems it may never even be noticed, provided that the array's *nominal operating voltage* (V_{no}) doesn't drop below that required to complete the charge cycle. The V_{no} is the maximum voltage produced when a cell, module, or array is functioning under actual field conditions.

Nominal operating voltage (Vno)**—voltage at a module's maximum power point under actual field conditions**

Many photovoltaic module manufacturers now list the output of their products at both STC and SOC. If only the rating at STC is available, the system designer must de-rate the modules from the ideal output taken at STC in order to size PV systems properly for actual field conditions of higher cell operating temperature and lower insolation. As we will see, the influence of cell operating temperature on module power production makes it well worth while to design the module and array mounting to allow for maximum free-air cooling of the array.

The I-V Curve

Each photovoltaic cell, module, and array has its own unique current/voltage relationship, known as an I-V curve, that can be graphically depicted as shown in figure 3-3. The graph shows the relationship between current (I) output, measured in amps, and voltage (V) output, measured in volts, from 0 volts at short-circuit current to 0 amperes at open-circuit voltage, as the resistance of the load across the module is increased from short to open circuit. *Short-circuit current* (I_{sc}) is current flowing unimpeded from its source through an external circuit with no load or resistance factor; it is the maximum flow of current (amperage) possible. *Open-circuit voltage* (V_{oc}) describes an open circuit with the absence of flowing current. The voltage potential across a photovoltaic cell (or module or array) in full sunlight in open circuit is the maximum possible voltage the device is capable of producing.

Isc**—short-circuit current**

Voc**—open-circuit voltage**

On every I-V curve, there is a point in the "knee" of the curve, roughly midway between short circuit and open circuit, at which the relationship of voltage to amperage is ideal—that is, where the product of the two, wattage, is the highest. This point is called the module's *peak power point* or *maximum power point*. The voltage at this point on the I-V curve is represented as V_m, the current as I_m, and output power or wattage as P_m.

Pm**—the power output in watts of an array or module at the maximum power point on an I-V curve**

Im**—current at the maximum power point on an I-V curve**

Vm**—voltage at the maximum power point on an I-V curve**

The sample I-V curve shown in figure 3-3, taken at STC of 1,000 W/m² and 25°C cell temperature, provides the following module performance data: I_{sc} of roughly 2.1 amps, V_{oc} of approximately 18.3 volts, I_m of slightly under 2 amps, and V_m of about 14.9 volts.

Module output current will vary linearly with solar intensity as shown by the family of I-V curves found in figure 3-4. For example, a 20 percent decrease in insolation results in a decrease in current output from 2.1 to 1.7 amps. Current also increases slightly as temperature rises. Voltage remains nearly constant as a function of solar intensity but is affected by variations in cell temperature, as shown in figure 3-5. As ambient

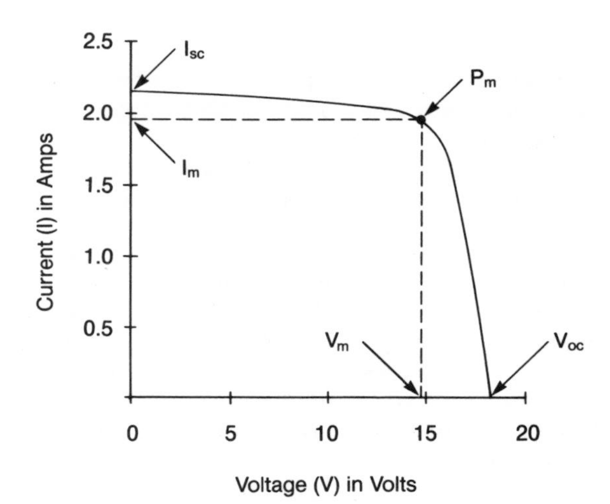

FIGURE 3-3

A sample I-V curve showing the relationship between current and voltage.

temperature rises, operating cell temperature also rises, reducing output voltage.

An I-V curve is used to make sure a photovoltaic system is operating as close as possible to its peak power point.

Peak power is at the knee of the I-V curve. It is the point at which the product of voltage and current is at its maximum. Figures 3-4 and 3-5 indicate that peak power is affected far more by variations in solar intensity than by variations in temperature.

I-V curves serve as useful design tools for a photovoltaic system designer and can be developed for cells, modules, panels, or arrays. They provide the necessary information to ensure that a PV system is configured to operate as close to its peak power point as possible.

The I-V curve is valuable both during the design phase and during installation and operation of a photovoltaic system. Through the use of an electronic device called an I-V curve tracer, it's possible to chart the course of a curve and its maximum power point either as part of a troubleshoot-

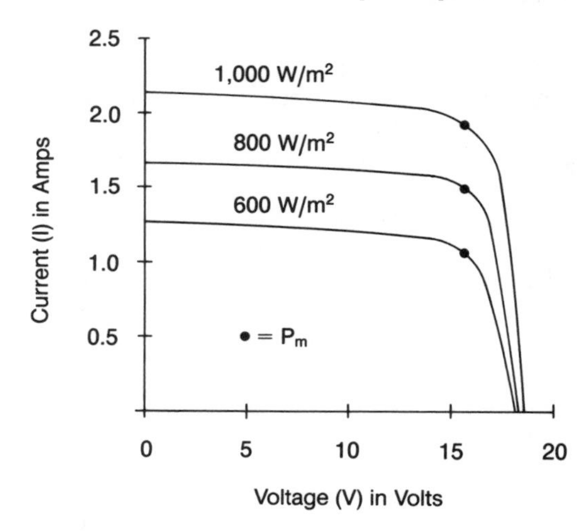

FIGURE 3-4

A module I-V curve showing the effect of insolation intensity on current.

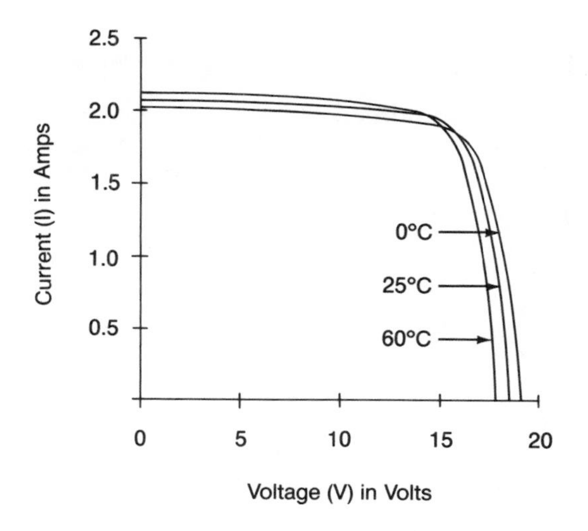

FIGURE 3-5
A module I-V curve showing the effect of cell temperature on voltage.

ing procedure or as a step in matching the modules to be connected in series to form an array string. Some manufacturers even provide I-V curve data with each new module to make this matching procedure that much simpler.

Many of today's larger and more sophisticated photovoltaic systems are designed to be used with an automatic electronic control circuit called a *maximum-power tracker* that forces the array to operate at its optimum power point. This allows the system to extract all the power available under constantly varying conditions. Max-power trackers, which will be discussed further in chapter 5, come as a standard feature in higher-quality DC-to-AC inverters and can also be purchased as discrete components for use in large-scale PV systems.

Comparing Photovoltaic Modules

Presently, there are several dozen manufacturers of commercial photovoltaic modules worldwide, many of whose products are well suited for use in residential installations. Table 3-1 compares a small sampling of these modules, representing a cross section of current approaches to silicon technology. Keep in mind as you read this table that innovative strides are made on an almost weekly basis in this dynamic new technology, so the specifications cited here necessarily reflect only a momentary state of the art.

In reading the table, remember that the voltage output for an individual photovoltaic cell remains at a fairly constant value of about 0.5 volt, give or take a fractional variation. Thus, the nominal 16.5-volt rating for the Mobil Solar Ra39-12 module represents an accumulated voltage made possible by the connection of a string of 36 cells in series. Connecting two of these 36-cell series strings in parallel to boost amperage has no effect on the voltage. The majority of the photovoltaic modules available on the

Table 3-1. A Comparison of Module Specifications

Manufacturer and Model	Cell Material	Size of Cells (in)	Number of Cells	Cell Connections	Operating Voltage (V)	Operating Current (A)	Peak Power (W)
ARCO Solar M-51	Czochralski silicon	4 (round)	35	All series	17.3	2.31	40
Mobil Solar Ra39-12	Ribbon silicon	2 × 4	72	36 series by 2 parallel	16.5	2.38	39
Kyocera PSA100H-361S	Polycrystalline silicon	4 × 4	36	All series	16.5	2.42	40
Mobil Solar Ra220-24	Ribbon silicon	2 × 4	432	72 series by 6 parallel	33.3	6.60	220
Mobil Solar Ra220-48	Ribbon silicon	2 × 4	432	144 series by 3 parallel	66.5	3.30	220
Solarex SX-42	Polycrystalline silicon	4 × 4	36	All series	16.5	2.60	43
ARCO Solar M-53	Czochralski silicon	4 × 4	36	All series	17.3	2.49	43
Solarex SX-146	Polycrystalline silicon	4 × 4	40	All series	18.0	2.60	47
Kyocera PSA100H-361H	Polycrystalline silicon	4 × 4	35	All series	16.5	2.67	44

market today are designed to charge 12-volt batteries. This means that they have the appropriate number of cells to deliver between 15.5 and 17 VDC under standard operating conditions.

For designers of large photovoltaic systems, the ideal PV module will have a nominal output voltage higher than the standard 12 volts—perhaps 24 or even 48 volts. Mobil Solar Energy Corporation was first to enter the marketplace with large-area, high-voltage modules. Their Ra-220 module, which measures 4 by 6 feet, can be ordered in 12-, 24-, 36-, or 48-volt versions to provide optimal flexibility to the system designer. These high-voltage modules are valued by the designer of large-scale PV systems because they require fewer module-to-module series string connections to deliver the high array voltages needed.

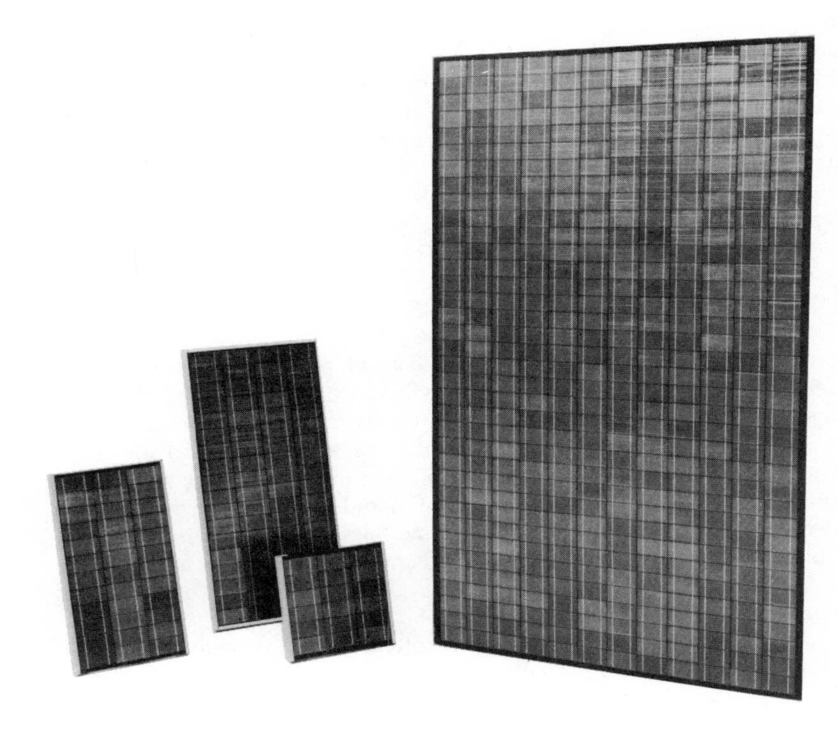

PHOTO 3-1
Mobil Solar Energy Corporation's family of silicon-ribbon photovoltaic modules, with the large-area Ra-220 shown at right.

A quick glance at table 3-1 might suggest that the module with the highest watt output represents the best value. This is not necessarily the case. A critical comparison must take into account cost in *dollars per peak watt.* The photovoltaics industry is extremely competitive, and manufacturers producing similar types of cells and modules have kept close pace with each other in dollar-per-watt costs. Buyers may assume, then, that the module's peak-watt output will be reflected in its price.

Price per peak watt, moreover, is not the only factor to consider when choosing a particular module. The manufacturer's basic standards of integrity and the long-term operational performance of the product are equally important. Residential array installations should be designed to meet the criteria set by the Jet Propulsion Lab. These criteria establish both electrical and mechanical design standards and measure the effects upon performance of environmental influences such as extreme temperatures, humidity, hailstones, wind, and snow loads. Photovoltaics enthusiasts who wish to review the JPL criteria should consult the U.S. Department of Energy publication, *Block IV Solar Cell Module Design and Test Specification for Residential Applications* (Jet Propulsion Laboratory, 1978).

In 1985, JPL announced it was discontinuing its work in support of the photovoltaics program because of DOE funding cutbacks. JPL has made a major contribution to the development of commercial photovoltaics and their participation will be sorely missed. The industry has adopted the JPL testing and rating criteria for PV modules, and it is expected that these standards will continue to be relied upon for the foreseeable future.

Modules are the basic building blocks of a photovoltaic system. Arriving at the correct number required to satisfy a given electrical load would seem to be a matter of simple arithmetic tempered by space and financial considerations. However, as we shall see, there is more to the sizing and design of a photovoltaic array than just establishing the total number of modules that will meet the load.

From Modules to Arrays

We know that a single photovoltaic cell produces approximately 0.5 volt at 2 amps. Since virtually every load we wish to serve requires higher levels of both voltage and current, every photovoltaic system is dependent upon an appropriate compounding of both voltage and amperage. This is accomplished at two levels: within the modules themselves and through the electrical interconnection of modules to form the array. Remember that connecting either cells or modules in series results in an accumulation of voltage, while connecting them in parallel causes amperage to be compounded. The difference between these two methods of connecting modules is shown in figures 3-6 and 3-7.

The voltage and amperage rating (and, by simple multiplication, the wattage rating) of a particular module is thus the result of the combination of series and parallel connections of individual cells within that module. We also know that these module output ratings are affected by operating conditions—most notably insolation and temperature. The understanding of these ratings, and the ability to configure multiple series/parallel combinations of modules to deliver the desired voltage and current, are prerequisite to the design of a photovoltaic array to handle a particular load requirement.

The Series String

The basic intermediary unit between module and array is the series string or source circuit. This is the number of individual modules con-

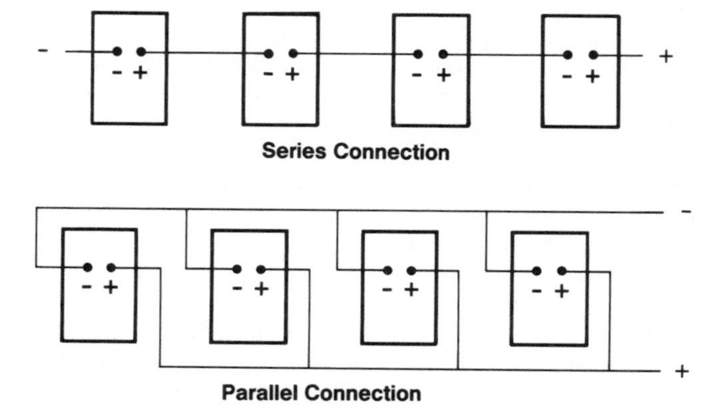

FIGURE 3-6
A diagram of four 12-volt, 2-amp photovoltaic modules wired in series, positive to negative, to deliver 48 volts at 2 amps.

FIGURE 3-7
A diagram of four 12-volt, 2-amp photovoltaic modules wired in parallel, positive to positive and negative to negative, to deliver 12 volts at 8 amps.

nected in series that will produce the nominal amount of voltage desired for the array output. The string itself thus becomes an integral electrical subcomponent of the array. The required number of series strings are then connected in parallel to bring the system amperage (and thus wattage) up to the desired level.

When assembling a series string, it is important to avoid or at least reduce module mismatch by joining together modules that are identical or nearly identical in performance (hence the usefulness of the I-V curve, discussed previously). This is because a series string, like a chain, is only as strong as its weakest link, that is, its weakest module or even its weakest cell. Despite the uniformity of testing procedures and manufacturers' claims for rated performance, the fact remains that output will vary some from module to module within a single product line. In short, the series string output will be no better than that of the poorest-performing module in the series. If the modules are purchased from a manufacturer who provides individual ratings, the job of matching them in a string is, of course, that much easier.

Series string—a group of photovoltaic modules wired together to form one series circuit; also known as an array string or source circuit

Mismatch is not nearly as much of a problem in the next phase of array assembly—the parallel connection of series strings to achieve the necessary accumulation of current. This is because a parallel sequence of connections, unlike a series, will result in an averaging of the individual strings' output rather than a reduction of output to the lowest common denominator. The stronger modules in a parallel string can help to pick up the slack for the weaker ones. The parallel connection of the individual series strings is usually done inside a device called a string combiner box, which will be described in chapter 4.

Evaluating the Solar Resource

A well-functioning photovoltaic power system is a partnership between PV hardware and the sun. Consequently, the attention given to the quality and availability of the solar resource is as important as that directed toward the choice of PV system components. Before we go further into the details of array sizing and design, we need to make certain we have a good understanding of the solar resource.

The output of a photovoltaic module or array is directly proportional to the amount of light that falls on its surface. As seasons change, so does the amount of available sunlight. As the position of the sun changes relative to the PV module, so does the amount of energy that the module can capture relative to the total energy that is available. The effect of sunlight intensity on PV system output can be shown with the help of an I-V curve (as in figure 3-4).

The amount of light available in a given area at a particular time of year can be calculated. But before we begin a detailed discussion of how to do this, let's first note a few special factors that must be taken into account:

- *The orientation of the module with respect to the sun.* Modules pointed directly at the sun will produce more power than those that are at an angle to the sun. Modules installed north of the equator are normally oriented toward true south (as opposed to magnetic south) and tilted at an angle from the horizontal that is within plus or minus 15 degrees of the site latitude.

- *Dirt accumulation on the front surface of the module.* Dirt absorbs sunlight so that it cannot reach the cells. In most areas normal rainfall is sufficient to keep modules relatively clean.
- *Shadowing by fixed objects.* Poles, antennas, chimneys, trees, or other tall structures that shadow an array during parts of the day will reduce its output. Shadowing from clouds and smog will also temporarily reduce output.
- *Reflection of sunlight.* Expanses of sand, snow, ice, and other light surfaces (including clouds, under the right circumstances) can increase output by reflecting additional insolation onto the modules.

Availability of Sunlight

An important factor in determining the total average daily solar energy falling on the earth's surface for a given month of the year is the number of hours of daylight. This is strictly a matter of the geometry of motion of the earth spinning about its axis once each day and traveling about the sun over the course of a year. Table 3-2 shows the number of hours of daylight on the second day of every other month at various latitudes within the Northern Hemisphere.

For any given latitude, the peculiarities of climate and weather at the specific site will determine the amount of solar energy that reaches the earth's surface. Phoenix, Arizona, and Charleston, South Carolina, are at roughly the same latitude and so see the same number of hours of sunshine on a clear day, but Phoenix receives approximately 30 percent more solar

Table 3-2. Daylight Hours as a Function of Latitude and Month of Year

Latitude and Location	Hours of Daylight per Day					
	Jan.	March	May	July	Sept.	Nov.
20°–30° Tampa; Houston; Mexico City	10.6	12.0	13.5	13.5	12.0	10.5
30°–40° Phoenix; Atlanta	10.0	12.0	14.0	14.0	12.0	9.9
40°–50° Boston; Omaha; Bucharest	9.1	12.0	14.8	14.8	12.0	8.9
50°–60° London; Berlin; Juneau	7.8	12.0	16.0	16.0	12.0	7.5

energy over the course of the year. This is because of the generally clearer atmosphere and fewer clouds in Phoenix. Climate conditions differ dramatically from site to site and must be carefully considered in the design of a photovoltaic system.

Because of this variation, statistical estimates of average daily solar energy levels, obtained from records taken at or near a site, must be relied on when designing a photovoltaic system. Sources for insolation data for the United States and international locations are listed in the bibliography.

Measuring the Sun's Power and Energy

Imagine for a moment that there is no atmosphere at all. You are in Phoenix, Arizona, it is January, and you want to know how much sunlight is available for your use. You decide to measure the power of the sun striking a 1-square-meter flat surface held perpendicular to the sun's rays at every instant from sunrise to sunset over the course of a single day. (Keeping the surface perpendicular to the sun's rays throughout the day as the sun moves is called *two-axis tracking of the sun.*) The amount of sun power, or solar insolation, impinging on the surface, measured in watts or kilowatts (kW) per square meter, would not vary from sunrise to sunset. It would remain at its value observed in space, in near-earth orbit, of 1.35 kW/m^2.

Two-axis tracking of the sun—a method of following the sun's daily path through the sky that keeps a flat surface continually perpendicular to the sun's rays

But consider the actual situation with the atmosphere present as we track the sun on a crisp, clear day. At solar noon, instead of 1.35 kilowatts per square meter, we measure only 1 kW/m^2 thanks to atmospheric reflection and absorption. (Because the sun's rays have more atmosphere to travel through near sunrise and sunset, the attenuation is more severe during early morning and late afternoon.) One kilowatt per square meter, then, is the nominal power of the sun at ground level on earth, the amount of radiation that impinges on a surface perpendicular to the sun at solar noon on a very clear day over a wide range of latitudes. It has become a universal reference, the basis of the definition of what constitutes *full sun power.*

Full sun power—1 kW/m^2, often referred to as one sun

Now let's say that instead of undertaking the complicated effort of holding a 1-square-meter surface perpendicular to the sun throughout every instant of the day, we simply leave it lying flat on a completely horizontal section of ground. The result is that, because of the slant of the rays, less power will strike this limited area than before, except when the sun is directly overhead.

One way to compensate for the reduction in power that occurs when the sun's rays move away from perpendicular is to enlarge the surface that absorbs the rays. There is a direct relationship between the angle of the rays and the additional surface area needed to capture solar power, as shown in figure 3-8. To determine the actual insolation value at any particular angle, multiply the full sun value at 90 degrees by the sine of the sun's present angle of incidence to the surface.

Once we take into account the effect of atmosphere on intensity of sunlight and the effect of angle to the surface on insolation level, we begin to get close to the actual conditions under which a solar electric system must function. So let's go back to Phoenix on a clear January day and see

FIGURE 3-8

A diagram showing the area required to capture equivalent solar energy as a function of surface tilt. (The smallest area is always perpendicular to the sun's rays.)

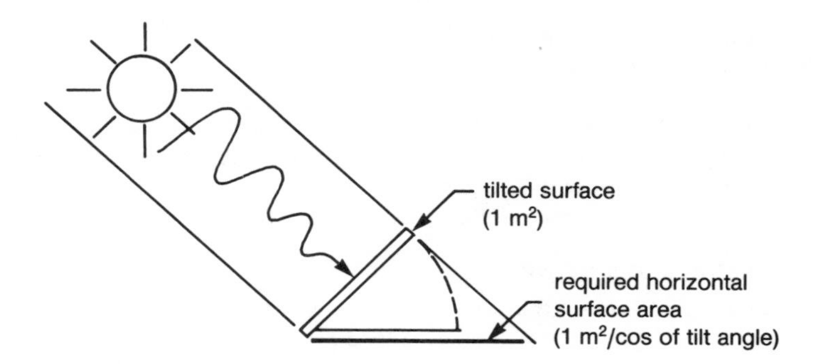

what we discover when we record the solar insolation that falls on a fixed horizontal surface during the course of a day. Figure 3-9 shows the results.

The most important thing we learn from these calculations is that even on a clear day, early in the morning and late in the afternoon we can expect little insolation on a horizontal surface. The effective power that reaches the surface is reduced because of the increased "thickness" of the atmosphere, which causes more light to be absorbed. Moreover, when the sun is at a low angle to a horizontal surface, a large portion of the sun's rays are simply reflected off the surface.

Equivalent Hours of Full Sun

Energy is the product of the intensity of a type of power and the duration of time that the power is made available. The total energy impinging on the square meter of horizontal surface during the course of the day is shown by the total area under the curve. In figure 3-9 this figures out to be 3.9 kilowatt-hours. This is, of course, less than what would be obtained if the surface were made to track the sun.

If the sun shone perpendicular to the surface at the full sun power level, which we have defined as 1 kW/m^2, it would have to shine on the surface for only 3.9 hours to produce the same result as a full day of "normal" sunlight striking the fixed surface. Thus, we can describe average daily insolation levels in terms of *equivalent hours of full sun*. The photovol-

FIGURE 3-9

A graph showing the relationship between power and time of day for insolation that strikes a fixed, horizontal surface in Phoenix on a clear day in January.

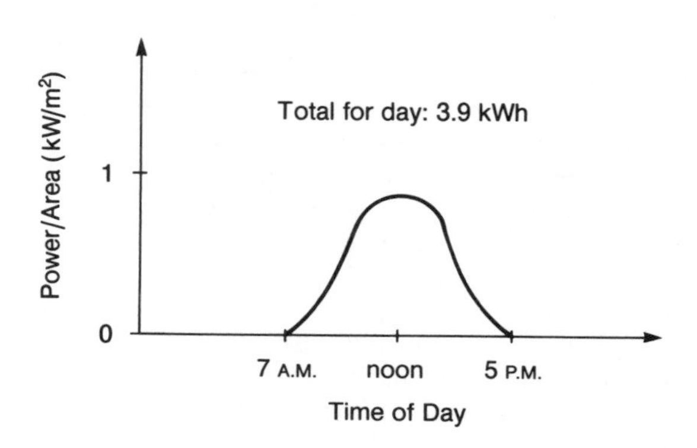

taic system design exercises found in this book consistently make use of this type of data.

When equivalent hours of full sun (often more briefly designated *full sun hours* or *peak sun hours*) are employed to measure solar insolation, it is always understood that the hours referred to are at the full sun intensity of 1 kilowatt per square meter. When calculated on a monthly basis, peak sun hour data takes into consideration not only latitude but also typical climatological conditions for the area in question. Thus, when this data is used as a basis for determining photovoltaic module output, no further correction for insolation levels at the site is necessary in the design calculations.

Equivalent hours of full sun—a conventional way of measuring the total amount of solar energy that impinges on 1 square meter of a fixed surface over the course of one day

Since full sun power is the nominal maximum insolation possible on the earth's surface, the number of equivalent hours of full sun is always much less than the actual number of hours from sunrise to sunset. So, for example, in Phoenix on a clear day in January, the actual duration of daylight is 10 hours. But the total energy impinging on 1 square meter of horizontal surface during those 10 hours can be described as equivalent to 3.9 full sun hours.

Of course, not every day is a clear day, even in Phoenix, Arizona. Because of haze, smog, clouds, rain, and other inhibiting factors, the average daily number of full sun hours will be less than the theoretical clear-day value in any month. For the month of January in Phoenix, the *average* daily value on a horizontal surface turns out to be 3.5 full sun hours, or 3.5 kilowatt-hours per square meter.

Not all site-specific variables are negative. Certain natural features—such as desert ground surfaces and expanses of snow, water, or ice—serve to reflect a significant amount of sunlight. This reflected light is as useful a resource for flat-plate photovoltaic modules as direct insolation. Because of these factors, a series of visits should be made to the site during as many different times of year as possible so that an accurate prediction of insolation availability can be made. If the site is extremely remote, or if time does not permit many or any site visits, an interpolation of recorded data for the nearest available site will have to suffice. When this is the case, and the availability of power at the installation is considered important or critical, the PV array should be designed with a margin of additional capacity to compensate for the lack of accurate data.

Effects of Array Orientation

So far we have discussed two ways to position a photovoltaic module. It can either be equipped with the mechanical means to pursue two-axis tracking of the sun, or it can be mounted at a fixed tilt equal to the horizontal. While it is clear that tracking makes possible a greater array output, the additional power gained usually does not justify the complexity and expense inherent in a tracking system. An exception may be made for very small PV systems used in the Sunbelt. For small arrays of three to six

Altitude angle—the angle formed between a horizontal surface on earth and a line connecting that point with the center of the sun

Azimuth angle—the angle of horizontal (east or west) deviation of the sun's position relative to true south

modules, a rather elegant passive solar tracker has been developed by Zomeworks that uses freon to provide reliable unattended polar-axis tracking. Despite the availability of such devices, the majority of PV systems in operation today utilize an array at a fixed tilt angle.

For optimum output, a photovoltaic module not only must be mounted at the proper tilt, it also must face in the best direction. For all locations in the United States, PV modules should be oriented toward true solar south for maximum output. The farther north the location of the installation, the more important the true south orientation is, because the path of the winter sun becomes ever more limited.

The sun's daily path changes throughout the year. The angle that is formed between the horizontal and an imaginary line that begins at solar noon from a particular location on earth and travels through the center of the sun is called the sun's *altitude angle*. This angle varies by over 46 degrees between the summer and winter solstice. Figures 3-10 and 3-11 show the relative positions of the sun's altitude angle in the summer and in the winter at solar noon.

When the path taken by the sun across the sky in relation to a fixed surface inclined at a particular angle is considered along with solar insolation data, the resultant relationship becomes a dynamic model. This is because the sun's angle of incidence to the fixed tilted surface is constantly changing as a compound function of both altitude angle and *azimuth angle* (the angle of deviation from true south).

Often the only insolation data recorded for many locations throughout the world is taken from a horizontal surface. This data is available on both a monthly and yearly basis and is found in documents like the Solar Energy Research Institute's *Insolation Data Manual* (U.S. Department of Energy, 1980). Of course, photovoltaic modules are not mounted horizontally (except perhaps at or near the equator), but instead are usually tilted at an angle from the horizontal corresponding to the site latitude. Thus, the amount of insolation striking this tilted surface will be somewhat greater than that which occurs at the horizontal. The formula for calculating the amount of *total tilted insolation* (TT) at a particular site using horizontal data is as follows:

$$TT = \frac{\text{Total horizontal insolation}}{\text{Cosine of (latitude} \times 0.85)}$$

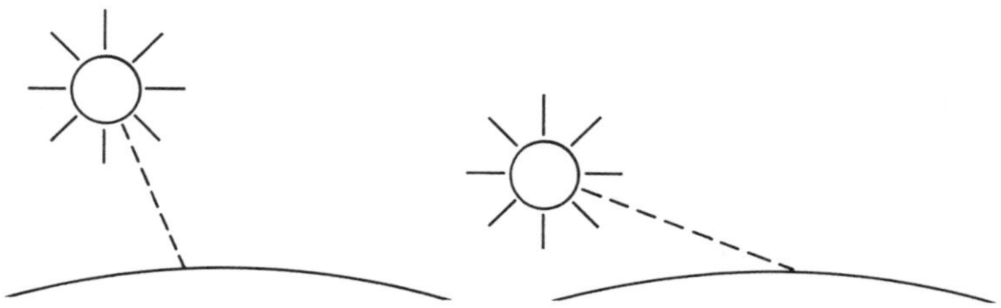

FIGURE 3-10
The altitude angle of the summer sun (left).

FIGURE 3-11
The altitude angle of the winter sun (right).

To demonstrate the formula, let's assume that our chosen photovoltaic installation site lies at 41° north latitude, in a location with an average yearly horizontal insolation of 3.5 kilowatt-hours per square meter. First, we multiply 41 degrees by 0.85 and get 34.85, which is equivalent to a latitude of 34°51' N. Next, using a calculator or trigonometry tables, we find the cosine of 34°51', which is 0.82065. Then we divide 3.5 by 0.82065 and get a result of 4.265. We thus learn that the total tilted insolation available at this spot is 4.265 kilowatt-hours per square meter. This is the annual daily average insolation striking a surface inclined at an altitude angle equal to the site latitude and facing due south. (If the site were in the Southern Hemisphere, it would face due north.)

Finding the Ideal Tilt Angle

Up till now, we have talked about using latitude to determine the best angle of tilt for a photovoltaic module. However, we have also noted that the altitude angle of the sun changes constantly over the course of a year. Setting our angle of tilt by latitude works fine midway between the solstices—that is, on March 21 and September 21—but is not ideal during the remainder of the year. During those other times, to find the ideal angle for our module we also would need to factor in the *declination angle*, that is, the number of degrees higher or lower than the latitude angle that the sun reaches at solar noon on a given day. This declination angle varies each day, changing from −23.45 degrees on December 21 to 0 degrees on March 21 to +23.45 degrees on June 21 and back to 0 degrees on September 21. The sun's altitude angle at solar noon for any given site on a given day can thus be expressed by the following simple formula:

$$\text{altitude angle} = 90° - \text{latitude angle} + \text{declination angle}$$

The problem, of course, is that the declination angle changes every day. Changing the tilt of an array consisting of more than a few modules on a daily basis often creates more problems than it is worth, and the increased electrical output that results seldom outweighs the labor costs involved.

***Declination angle*—the angular position of the sun at its highest point in the sky with respect to the plane of the equator**

In actual practice, the tilt of a flat-plate photovoltaic array should be changed a minimum number of times over the course of a year, if at all. Generally speaking, it is best to determine the optimum fixed tilt for the system being designed and size the array accordingly. This produces the best overall results in terms of cost and complexity of the system.

For a load that remains constant throughout the year, the best approach is to set the tilt angle equal to the latitude plus 15 degrees, which orients the array toward the low altitude angle of the sun in winter when the least amount of insolation is available. Even though this is about 30 degrees away from the best "summer tilt," the extra light available from the sun during the summer makes up for this nonoptimum setting. The result is a fairly steady module output throughout the year.

The tables that follow provide typical representative insolation data for three American cities: Phoenix (table 3-3), Boston (table 3-4), and Albuquerque (table 3-5). In each case the average amount of solar energy

Table 3-3. Insolation Values at Various Tilt Angles for Phoenix, Arizona

Tilt Angle	Average Energy Available (kWh/m²/day) Jan.	Feb.	March	April	May	June	July	Aug.	Sept.	Oct.	Nov.	Dec.	Year Total	Average
Horizontal	3.6	4.5	6.0	7.5	8.4	8.1	7.7	6.9	6.1	4.9	3.8	3.1	70.6	5.9
Latitude −15°	5.0	5.7	6.9	7.8	8.2	7.7	7.4	6.9	6.6	5.8	5.0	4.3	77.3	6.4
Latitude	5.8	6.3	7.2	7.7	7.6	7.0	6.8	6.6	6.7	6.3	5.7	5.0	78.7	6.6
Latitude +15°	6.2	6.5	7.1	7.1	6.6	5.9	5.8	5.9	6.3	6.4	6.0	5.4	75.2	6.3

Table 3-4. Insolation Values at Various Tilt Angles for Boston, Massachusetts

Tilt Angle	Average Energy Available (kWh/m²/day) Jan.	Feb.	March	April	May	June	July	Aug.	Sept.	Oct.	Nov.	Dec.	Year Total	Average
Horizontal	1.8	2.5	3.8	4.5	5.6	5.9	5.7	5.2	4.2	2.8	1.8	1.6	45.4	3.8
Latitude −15°	2.7	3.3	4.5	4.8	5.5	5.7	5.5	5.3	4.7	3.5	2.5	2.5	50.5	4.2
Latitude	3.1	3.6	4.6	4.6	5.2	5.2	5.1	5.0	4.7	3.6	2.7	2.8	50.2	4.2
Latitude +15°	3.2	3.6	4.5	4.2	4.6	4.5	4.5	4.5	4.4	3.6	2.7	3.0	47.3	3.9

Table 3-5. Insolation Values at Various Tilt Angles for Albuquerque, New Mexico

Tilt Angle	Average Energy Available (kWh/m²/day) Jan.	Feb.	March	April	May	June	July	Aug.	Sept.	Oct.	Nov.	Dec.	Year Total	Average
Horizontal	3.6	4.5	5.9	7.3	8.3	8.6	8.1	7.6	6.2	5.1	3.8	3.2	72.2	6.0
Latitude −15°	5.0	5.9	6.9	7.7	8.1	8.2	7.8	7.7	6.8	6.3	5.2	4.7	80.5	6.7
Latitude	6.1	6.5	7.2	7.5	7.5	7.4	7.1	7.3	6.9	6.8	5.9	5.6	81.8	6.8
Latitude +15°	6.7	6.8	7.1	6.9	6.6	6.2	6.1	6.6	6.5	6.9	6.3	6.1	78.8	6.6

available on a 1-square-meter surface during a single day of each month is presented in terms of full sun hours (total kilowatt-hours per day). Notice the variation in the relative amount of insolation at different times of the year depending on whether the surface is horizontal or tilted to latitude minus 15 degrees, to latitude alone, or to latitude plus 15 degrees.

You will note that in each of the three locations December is the *worst-case month,* the month with the least average insolation. Looking at the data in each column you will see that in December, the best tilt angle for the photovoltaic array is always latitude plus 15 degrees. Since a PV system is able to put out at least the worst month's potential power throughout the year, it is often designed around the conditions that exist during that month.

Very often, however, the photovoltaic system designer has to consider other factors when evaluating the optimum orientation for a particular array. For example, there may be certain microclimatic conditions that are specific to the site, such as early-morning fog or predominant afternoon cloudiness. Or the slope of an existing roof on which the array is to be mounted may determine the choice. A number of variables may influence the final decision as to the best angle for mounting the array.

Photovoltaic Array Sizing and Design

The process of sizing a photovoltaic array begins by determining the number of PV modules that are needed to ensure that the user of the system can rely on receiving the desired amount of electrical power. To accomplish this, the PV system designer requires detailed information on two basic parameters: first, the amount of electrical power required by the load; and second, the amount of solar energy available at the site.

Performing a detailed design analysis for a photovoltaic system requires accurate knowledge of the load in terms of both the amount of electricity required *and* the time when it is required. This is easy to calculate for a "fixed load" such as a radio repeater or a water pump that will operate X hours per day and requires Y watts of power per hour of operation. Calculations are more difficult for a PV-powered residence where the owners' daily routine changes frequently, guests visit, and additional loads are connected to the system after installation.

Since chapter 6 is devoted to the subject of residential load calculations and chapters 7 and 8 provide detailed illustrations on methods of sizing photovoltaic arrays, we will limit ourselves here to simple examples that illustrate the principles involved in array sizing.

To begin with, we should note that there are two basic methods commonly employed in the sizing of PV arrays. The first method utilizes calculations whose electrical units are expressed in terms of ampere-hours, or amp-hours. This method is most often used in the sizing of stand-alone (SA) systems that are used to charge a battery storage bank. The second method of array sizing expresses electrical units in terms of watt-hours or kilowatt-hours and is usually employed when sizing utility-interactive (UI) photovoltaic arrays or SA systems that work without battery storage (such as direct-coupled water pumping systems). We will examine each method in turn.

Array Sizing in Terms of Ampere-Hours

In the ampere-hour method of array sizing, the nominal output voltage of the array is configured (by connecting modules in series) to match the voltage requirements of the load. The peak power current of the photovoltaic module(s) or strings of series modules (expressed in amps under "full sun" of 1 kW/m^2) is then multiplied by the average number of peak sun hours per day obtained from a data base on insolation for the chosen altitude angle at the site during the month or season of highest load. The result is the average number of amp-hours per day of output that can be expected from the array.

When we are sizing low-voltage battery-charging photovoltaic systems—those less than 48 volts—the amp-hour method of array sizing usually allows us to ignore the fluctuations in voltage that occur in the array output due to variations in temperature and load. This can be done because almost all PV modules are designed for battery charging. This means that the module's nominal operating voltage is approximately 25 percent higher than the battery's nominal rated voltage. For example, a 12-volt nominal PV module has an average "working voltage" of 15 to 17 volts. An array voltage that has been configured to match the battery's nominal operating voltage must always be higher than the battery voltage for charging to occur.

The ampere-hour method of array sizing is popular because it relates well with the common method of sizing a battery storage bank, which is also based on amp-hours. The amp-hour sizing principle also automatically takes battery efficiency into account so that, for the purposes of our array sizing calculations, the battery can be considered 100 percent efficient in terms of *amp-hours in versus amp-hours out*. Batteries and battery sizing will be discussed in detail in chapter 4.

The array is usually designed to satisfy the average daily load for the worst-case month of the year. Once that load has been calculated and the amount of available sun at the site has been determined, a safety factor is introduced to account for degradation in array output due to dirt accumulation, internal losses from wire runs and system components, and general uncertainties about "average" outputs and inputs. This safety factor should also provide the system with enough additional capacity to allow the array to service the daily load while simultaneously recharging the battery after a long period of no sun. The safety factor is usually between 10 and 20 percent, depending on how critical the load is and how conservative you want to be.

In sum, the principal parameters that need to be determined in order to size the photovoltaic array in a small SA system are the load on the system (expressed in average daily or peak daily amp-hours) and the amount of solar energy available at the project site (expressed in average daily peak sun hours per month at the altitude angle above the horizontal that corresponds to your chosen array mounting angle).

Example A: One Very Basic System

Here is a simplified example that uses a very basic SA system to illustrate the fundamental principles of the ampere-hour method of photovoltaic array sizing. A PV system is required to power one 20-watt fluorescent lamp with DC ballast in a remote cabin located in central Massachusetts. The fixture's power requirements are 1.8 amps at 12 volts (nominal) DC. An average of 4 hours of light per day are required seven days per week during late summer and early fall (August, September, and October). Availability of power for this small load is not considered critical.

With these facts in mind, here are the steps to follow in sizing the array:

Step 1: Determine the average daily load by multiplying the load current in amps (A) by the average hours of daily use (h). In this case, we get a result of 7.2 amp-hrs (1.8 A × 4 h = 7.2 Ah).

Step 2: Determine the full sun or peak sun hour value to be used in the design. Since output will be required from this system only during late summer and early fall, we select an array altitude mounting angle that is equal to site latitude. By referring to table 3-4 we learn the number of equivalent hours of full sun on a surface tilted at an angle equal to the latitude for the Boston area during the months of August, September, and October. The figures are 5 hours (August), 4.7 hours (September), and 3.6 hours (October).

Since this is a noncritical system, we can size the array according to the average full sun figure, which is 4.4 hours. Had this been a critical application, we would have sized for the worst-case condition of 3.6 full sun hours and also provided an extra margin of battery storage capacity.

Step 3: Determine the required array size. The array charges the 12-volt battery on sunny days when output exceeds the load. On cloudy days and certainly at night, the load will exceed the array output and drain the battery. The array must then be sized to ensure that over the long run the balance is positive and the battery is more likely to be charged than discharged.

The array must be sized to deliver an average daily output equal to the average daily system load (including all internal losses) plus approximately 10 percent to ensure that the battery will be recharged *and* that the load will be served after a long period of "no-sun" days. If this is not done and the array output is set to just meet the load demand, the battery will never regain its charge from a discharged condition; after the daily load requirements are met, there will be no surplus left for the battery.

This slight oversizing of the array is good design practice. However, since this is a simple system and the load is considered noncritical, one can argue that the cabin's owner could just get along with a little less light for a while to allow the battery to regain its charge after a worst-case

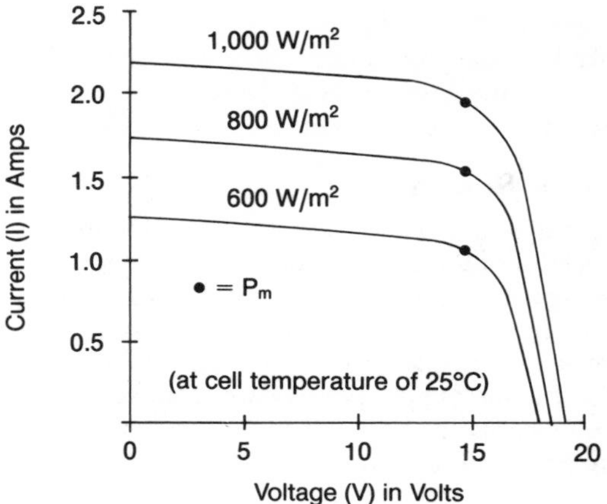

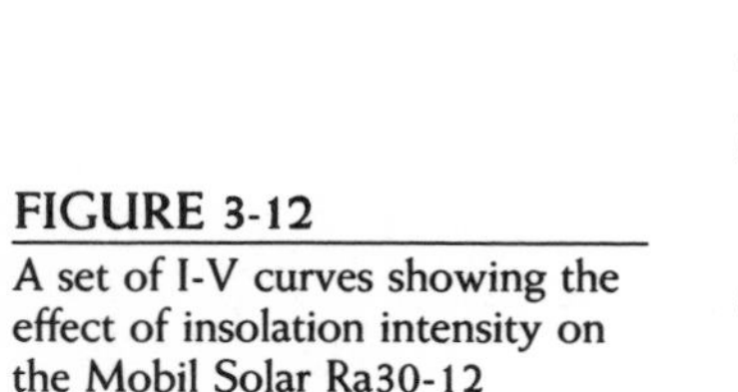

FIGURE 3-12

A set of I-V curves showing the effect of insolation intensity on the Mobil Solar Ra30-12 module.

period of no-sun days. After all, the need for this slight inconvenience will arise infrequently.

Let's assume that in this case the owner understands these parameters and desires to save money, so we will not add the extra percentage to the array sizing. The array will simply be sized to meet the average daily load, which must include all the internal system losses in the wire runs and electronic controls.

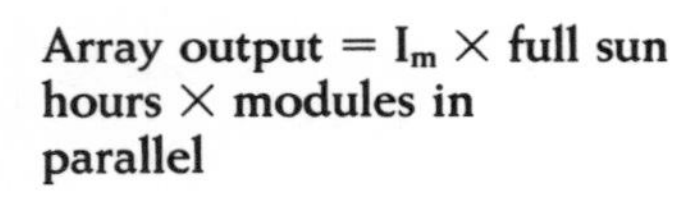

Array output = I_m × full sun hours × modules in parallel

The average daily output of a photovoltaic array, calculated in terms of amp-hours delivered at the array's nominal voltage, is the product of the number of (average) full sun hours times the actual peak power current (I_m) at NOCT of one of the modules in the array times the number of modules in parallel. Or, to state the matter in a briefer, formulaic manner: array output = I_m × full sun hours × modules in parallel.

As we continue with our efforts to size a photovoltaic array for our central Massachusetts cabin, let's assume that we will design the system to make use of the Ra30-12 module manufactured by Mobil Solar Energy Corporation. A set of I-V curves for this module are provided in figure

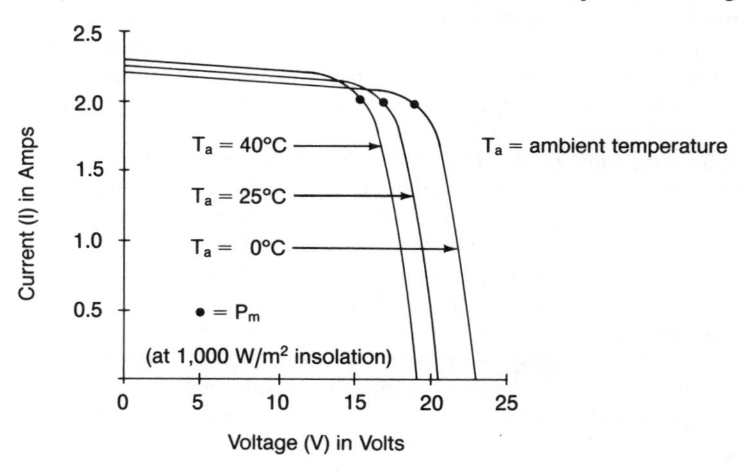

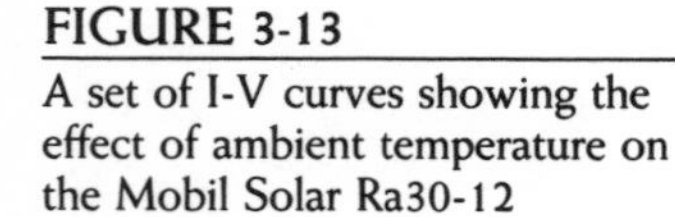

FIGURE 3-13

A set of I-V curves showing the effect of ambient temperature on the Mobil Solar Ra30-12 module.

3-12. They illustrate the dependence of the module output current on insolation intensity. Figure 3-13 presents a set of I-V curves that demonstrate the effects of ambient temperature on the same module.

The Ra30-12 module is rated at 30 peak watts and is designed to charge a 12-volt battery. For our system's site, which has an average daily ambient temperature of about 10°C (50°F), the module's peak power current is 2 amps under full sun conditions. The average daily equivalent hours of full sun at this site is 4.4 hours. Multiplying that figure by the I_m of 2, we determine the output of a single module is 8.8 amp-hours per day.

Step 4: Select the number and type of modules. In Step 1, the average daily load was determined to be 7.2 amp-hours. By delivering 8.8 amp-hours a day, a single Ra30-12 module will satisfy this load with some power to spare. This "surplus" power will be enough to cover the internal losses in our system from the voltage regulator and wire runs.

As we noted earlier, batteries are considered 100 percent efficient in terms of amp-hours in versus amp-hours out. What actually happens is that the amp-hours are put into the battery bank at a higher voltage than the voltage at which they are taken out: charging takes place at 13 to 15 volts, discharging at 11 to 13 volts. Thus, the efficiency loss in the batteries is compensated for by the PV module's higher operating voltage.

Using a single Ra30 module, the owner of the remote cabin in central Massachusetts will also get his extra margin of array output, so that the battery in his system will regain its charge after a long period of no-sun days without any interruption in service. In short, we conclude that this person can get by with an array size of one 12-volt, 30-peak-watt module.

That seemed rather easy, didn't it? Well, it was. But then it was just a simple example chosen to present the basics. So let's look now at another hypothetical example, one that is slightly more complex, requires configuring modules in series and parallel, and takes seasonal load variations into account.

Example B: A More Complex System

In this case, a 48-volt photovoltaic system is required to power a remote retreat cabin in the foothills of the Sierra Nevada mountains in northern California. The site elevation is 5,000 feet and the latitude is 40°N. The owner plans to use the cabin for a three-week vacation each summer and for three-day weekends in winter.

Consulting a resource on insolation data, we find that the site has an average of 3 peak sun hours in the winter months and 5.5 peak sun hours in the summer months on a surface inclined at an altitude angle equal to latitude. Table 3-6 shows the owner's estimated power requirements for each electrical load in the cabin and the average daily hours of use.

After an initial review of the seasonal loads and the owner's use patterns, we will assume for our initial design that the photovoltaic array will be mounted at a 40-degree (latitude) tilt all year to help balance winter output and load with summer output and load. We will then have 5.5 hours peak output in the summer and 3 hours peak output in the winter.

First, let's calculate the average daily winter load in ampere-hours

Table 3-6. Average Daily Loads for a Remote Cabin in California

Loads	Amperage (A)	Summer Hours	Winter Hours
48 VDC			
4 fluorescents (40 W)	1.0 ea.	2.0	5.0
6 fluorescents (20 W)	0.4 ea.	2.0	4.0
Refrigerator (8 cu ft)	1.5	12.0	10.0
Well pump	11.0	0.7	0.5
120 VAC			
CB radio	0.2	1.0	1.0
17″ color TV	0.5	3.0	3.0
Miscellaneous appliances	2.0	0.5	0.5
Small washing machine	4.0	0.5	0.0
Stereo system	0.5	2.0	3.0

by multiplying the number of amps needed by each item by the average daily number of hours that the item is in operation. We can then use the results of these calculations to determine the required array size. Here are the figures for the 48-VDC loads:

lights (4 @ 40 W)	4.0 A × 5.0 hr =	20.0 Ah
lights (6 @ 20 W)	1.6 A × 4.0 hr =	9.6 Ah
refrigerator (8 cu ft)	1.5 A × 10.0 hr =	15.0 Ah
well pump	11.0 A × 0.5 hr =	5.5 Ah
	Total	50.1 Ah/day of use (48 VDC)

The figures for the 120-VAC loads look like this:

CB radio	0.2 A × 1.0 hr =	0.2 Ah
17″ color TV	0.5 A × 3.0 hr =	1.5 Ah
miscellaneous appliances	2.0 A × 0.5 hr =	1.0 Ah
small washing machine	4.0 A × 0.0 hr =	0.0 Ah
stereo system	0.5 A × 3.0 hr =	1.5 Ah
	Total	4.2 Ah/day of use (120 VAC)

An inverter will be needed to furnish the 120-VAC supply. So we choose a high-efficiency 2-kilowatt inverter with a 48-VDC input. We assume an *average* efficiency of 75 percent. The assumed internal loss of 25 percent of the power that enters the inverter must then be added to the 120-VAC load requirements so that we can find the average daily winter load on the system at 48 VDC.

In making the conversion from the 48-VDC to the 120-VAC power, keep in mind the rule that power equals voltage times amperage. As voltage goes up, amperage goes down accordingly. Thus, we first divide our AC voltage (120) by our DC voltage (48). From this we learn that we will have to generate 2.5 times as many ampere-hours of DC current as we need of AC current (in this case, 4.2 amp-hours). Thus it would appear that we need to generate 10.5 amp-hours. However, since we are figuring a 25 percent loss of power during the conversion process, we must divide this figure by 0.75. Doing this we find that we really must generate 14 amp-hours. Finally, we add the 14 amp-hours of direct current needed to cover the AC load to the 50.1 amp-hours we previously calculated for the straight DC load and conclude that our total daily winter load is 64.1 amp-hours at 48 VDC.

However, we still have not completed our calculations, since this cabin will be used for a maximum of three days a week during the winter. Assuming that power is needed three out of every seven days it is generated, we multiply 64.1 by 0.43 (the decimal equivalent of 3/7) and find that we have only 27.6 amp-hours of use per *average day* in the winter. Then, figuring 10 percent internal system losses for wiring runs and the voltage regulator, we divide 27.6 by 0.90 and conclude that we need 30.66 amp-hours per average day of charging current from the array to satisfy the typical winter load demand.

The next step in the system sizing process is to factor in the available amount of sunlight. Earlier we noted that at the chosen site and with the array tilted to the desired angle, we can count on 3 hours of peak sun per day during the winter months. When we divide our average daily requirement of 30.66 amp-hours by the number of average daily full sun hours (3), we find that we need an array capable of generating 10.22 amps. Mobil Solar's Ra30-12 module delivers an output of 2 amps under full sun. If we choose to use this particular module in our system we see that we need 5.11 modules connected in parallel to produce the needed amperage.

Obviously we cannot make use of a fraction of a module in our system, so we'll round the figure up to the next highest number. This will provide a margin of additional capacity to account for long periods of poor winter weather conditions. We thus find that 6 Ra30-12 modules joined in parallel will provide the current needed to satisfy the winter load in this application. In addition, 4 modules, at 12 volts each, joined in series are needed to deliver the required nominal 48 volts for the system. Multiplying the 4 modules in series by the 6 series strings in parallel, we end up with a total array of 24 modules, each 12 volts and 30 peak watts.

We have demonstrated how the average daily number of ampere-hours can be determined for the winter load at this cabin, and with this information we have calculated the array size required to meet the load.

Now let's go through the same series of calculations employing the data we are given for the summer load. First let's consider the 48-VDC loads:

lights (4 @ 40 W)	4.0 A × 2.0 hr =	8.0 Ah
lights (6 @ 20 W)	1.6 A × 2.0 hr =	4.8 Ah
refrigerator (8 cu ft)	1.5 A × 12.0 hr =	8.0 Ah
well pump	11.0 A × 0.7 hr =	7.7 Ah
	Total	38.5 Ah/day of use (48 VDC)

The figures for the 120-VAC loads look like this:

CB radio	0.2 A × 1.0 hr =	0.2 Ah
17″ color TV	0.5 A × 3.0 hr =	1.5 Ah
miscellaneous appliances	2.0 A × 0.5 hr =	1.0 Ah
small washing machine	4.0 A × 0.5 hr =	2.0 Ah
stereo system	0.5 A × 2.0 hr =	1.0 Ah
	Total	5.7 Ah/day of use (120 VAC)

We must again account for the inverter losses and will continue to assume an average efficiency of 75 percent. This internal loss must be added to the 120-VAC load requirements to find the average daily summer load on the 48-volt battery bank. Here is our calculation:

$$(120\ \text{V} \div 48\ \text{V}) \times (5.7\ \text{Ah} \div 0.75) = 19\ \text{Ah @ } 48\ \text{V}$$

When we add the 19 amp-hours needed to cover the AC load to the 38.5 amp-hours needed for the straight DC load, we see that our average daily summer load is 57.5 amp-hours at 48 volts.

The cabin will have continuous summertime use during the three-week vacation period and therefore will consume 57.5 ampere-hours per day. Employing the same 10 percent figure for internal system losses, we revise this number to 63.9 amp-hours of array output per average day. Thus, at 5.5 hours per day of average peak summer sunlight, we need 11.62 amps (63.9 ÷ 5.5) of output from the photovoltaic array. We know each module is capable of producing 2 amps per day, so we divide 11.62 by 2 and find that 5.81 modules are necessary to generate the needed amperage.

Once again we round up to the next even number and gain an extra margin of capacity. From these basic calculations, it can be seen that the summer requirement of 63.9 amp-hours per average day is the higher of the two requirements and will be the chosen design value. In this case, though, the result ends up the same. For both summer and winter we conclude that we need an array of 6 series strings of 4 modules each—or 24 modules at 12 volts and 30 peak watts each—to satisfy the load requirement.

If this were an actual system design exercise, at this point in the process the conscientious systems designer would review his calculations with the owner to confirm the loads and their daily duration. Possibly they would then discuss various conservation options that could reduce the size and cost of the system. These options might include substituting a gas-fired refrigerator for the DC-powered model, further reducing the lighting load, carefully reviewing the well pump for possible improvements in efficiency, reducing or eliminating the inverter operation, and so on. For the purpose of this example, we will consider the load as fixed. Thus, the array for this system will be sized at 24 modules at 30 peak watts for a total of 720 peak watts of capacity.

Array Sizing in Terms of Watt-Hours

We have reviewed the steps necessary for sizing a photovoltaic system in terms of amp-hours. Now let's consider how a system can be sized in terms of watt-hours. Let's take as our example a larger PV array that has many modules wired in series and delivers a higher DC voltage. In order to determine the array voltage needed for a particular system and the number of modules required in series to deliver this voltage, we must first establish the nominal DC voltage (or voltage range) required by the load. Once this is done, the voltage rating of the PV module or array must be calculated based on the temperature and sun conditions that it will be working under in the field. A nominal operating cell temperature must be established, along with an expected average insolation level. The NOCT will influence the module's nominal operating voltage (V_{no}) and, by extension, its power output (W_p).

The influence of operating temperature on array-sizing calculations becomes increasingly important as system size and string voltage increase. This influence will have a direct impact on the number of modules needed to form the series string required to deliver the DC output voltage desired from the array. The average insolation level will influence the module's output current, which will directly affect the number of strings required in parallel to satisfy the current requirements of the load.

The term *nominal* is often used in connection with the voltage of a photovoltaic module or array because, as noted previously, the voltage of a PV device varies under different operating conditions—most notably, temperature and load. The open-circuit voltage of a PV module declines by approximately 0.5 percent per degree Celsius of temperature rise. The variation in output voltage will follow the module's I-V curve as a function of the resistance of the load. It will range from little or no voltage (and high current) at very low resistance to near the highest open-circuit voltage (with little or no current) at very high resistance. Those familiar with electricity will recognize this relationship as *Ohm's law,* which simply states that the voltage developed in an electrical circuit is equal to the value of the current in amps multiplied by the value of the resistance in ohms.

***Ohm's law*—voltage in an electrical circuit is equal to the value of the current in amps times the value of the resistance in ohms**

The influence of the load on photovoltaic output voltage can be easily illustrated using a simple PV system designed to charge a 12-volt (nominal) battery bank as an example. This PV array will operate at anywhere from 10 to 20 volts independent of temperature influences. The lowest array voltage will occur when the battery is fully discharged. The battery's own voltage is low and it wants and can accept a large amount of current. As the battery begins to regain its charge, its voltage increases. In response to this, the operating voltage of the PV array also increases and the array current must then decrease.

This process continues until the battery is fully recharged. At that point the PV array will be operating at approximately 15 to 16 volts. The system's voltage regulator will terminate the charging procedure and disconnect the array from the battery. Current will cease to flow, and the array voltage will rise to its maximum, the open-circuit voltage, which is approximately 20 volts for most 12-volt (nominal) PV modules. Voltage excursions in a PV system designed to charge a 24-volt battery bank would be from 20 to 40 volts. In a 48-volt system, it would be 40 to 80 volts.

The phenomenon of open-circuit voltage affects the design process, since components—power-conditioning equipment in particular—must be able to withstand transient periods of increased voltage. We are not talking here about super-high-voltage lightning transients, which in any case are handled by the surge-protection devices called *varistors* that are an essential part of any PV system. Rather, we are concerned with current and voltage boosts associated with temporary variations in insolation and temperature.

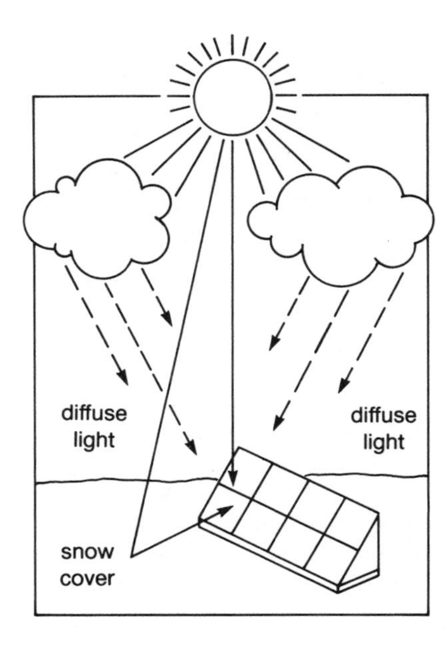

FIGURE 3-14

The earth's atmosphere, clouds, and ground reflectants influence the amount of sunlight that strikes a photovoltaic array.

For example, if, during a period of uniformly distributed diffuse insolation across an array, a strong path of clear, direct sunlight radiates through a sudden opening in the clouds (as shown in figure 3-14) and strikes a portion of that array, current will increase markedly. This is particularly the case if there are other favorable variables present, such as a strongly reflective snow cover. (The occurrence of 1.5 times normal peak insolation and more has been recorded.) The situation is exacerbated if temperatures are low, favoring increased voltage, and if the system is in or approaching a state of open-circuit voltage, with little or no current flowing to loads. These short-lived "surges" do little to boost kilowatt-hour output but can severely stress electronic components unless a margin of safety is built into the design.

Understanding all the sources and implications of voltage variations in photovoltaic array output is fundamental to good PV system design, especially when high-voltage arrays (above 48 VDC) are being designed. This is because the voltage variations and their impact on the system become greater as the system size and voltage increase.

As we saw in the first example, the choices are simple when you are working with low-voltage systems. A 12-volt system will employ one 12-volt (nominal) module in series by however many are required in parallel to deliver the required current. A 24-volt system would, of course, have two 12-volt modules in series. However, when you are designing a larger, higher-voltage PV array, such as that required to deliver 240 VDC (nominal) input to a UI inverter or a large battery bank, things get a bit more complicated and the effects of temperature and insolation must be more carefully considered. Let's look at such an array in order to study the influence of cell operating temperature on PV array output voltage.

Sizing a Large, High-Voltage Photovoltaic Array

For our example this time, let's assume we face the task of designing a 4.5-peak-kilowatt UI system where the photovoltaic array will deliver a nominal 240 VDC to the power conditioner, the electronic hardware that converts the array's DC power output to AC electricity.

For this example we will work with the criteria previously introduced as standard operating conditions. These include insolation of 800 W/m^2, an average air temperature of 20°C (68°F), a wind velocity of 1 meter per second, and a NOCT of 20 to 40 degrees above ambient temperature. (The relationship of NOCT to ambient temperature is influenced by insolation levels, because of different degrees of infrared gain, and variations in the amount of heat dissipated because of the method of module mounting.)

The range of possible variation in ambient temperature at our hypothetical site will be from −20° to 38°C (−5° to 100°F). In this case we will assume a NOCT 25 degrees above ambient, giving us a variation in operating cell temperature of 5° to 63°C (41° to 145°F). The average NOCT will be assumed to be 46°C (115°F). With this range in the NOCT, the open-circuit voltage of a series string of 20 photovoltaic modules at 12 volts (nominal) would vary from over 430 volts during cold winter conditions to less than 340 volts during the hottest summer weather, independent of insolation influences. In a system of this size, the range in operating cell temperature produces a rather large variation in output (especially voltage), which cannot be simply overlooked as in the sizing examples that were presented earlier for the low-voltage, battery-charging systems.

To begin our design effort, we must first review the input voltage requirements of the power conditioner to make certain that the array is designed to function most efficiently. We find that the power-conditioning unit (PCU) chosen for this example, a model UI-4000 made by American Power Conversion Corporation, has the following input voltage specifications:

- Nominal input voltage and current: 230 VDC @ 19 amps
- Full-power operating range: 200 to 300 VDC input
- Minimum start-up voltage: 231 VDC input
- Maximum start-up voltage: 338 VDC input

These specifications define a "voltage window," high and low values for the DC input voltage from the array, outside of which the unit will not function. The array must be designed to satisfy these input requirements during the majority of its operating hours. In UI systems, the efficiency losses internal to the PCU, which are generally less than 6 to 10 percent, are not usually taken into consideration in array sizing per se; rather they are viewed as part of the load.

At the outset we must establish the number of modules required in series for an array string that will operate within the voltage window established by the requirements of the PCU. If we again use the Mobil Solar Ra30-12 module, we find from the manufacturer's literature that under a NOCT of 46°C it has a nominal operating voltage of 13.5 volts at its maximum power point and an open-circuit voltage of 17 volts.

We determined earlier that the array's nominal operating cell temperature could vary from 5° to 63°C at this site. We also know that a module's voltage will decrease 0.5 percent per degree Celsius of temperature rise. Using the module's V_{no} of 13.5 volts taken at a NOCT of 46°C as a base, we discover the following:

$$1 - [(63° - 46°C) \times .005\ V/°C] \times 13.5\ V = 12.35\ V$$

$$1 + [(46° - 5°C) \times .005\ V/°C] \times 13.5\ V = 16.27\ V$$

That is, we find we have a high V_{no} of 16.27 volts and a low V_{no} of 12.35 volts under extreme conditions. Knowing we must fit within the load's voltage window, we try 18 modules in a series string to see what they will give us.

$$18 \text{ modules} \times 16.27\ V/\text{module} = 292.86\ V\ (\text{high})$$

$$18 \text{ modules} \times 12.35\ V/\text{module} = 222.30\ V\ (\text{low})$$

We see that the output of this series string fits very nicely within the required voltage window.

Next we check the module's open-circuit voltage. Starting with the V_{oc} of 17 volts taken at a NOCT of 46°C and using the same procedure as with V_{no} we find that the V_{oc} will vary from a low of 15.55 to a high of 20.49 volts. Checking this against the PCU's maximum start-up voltage of 338, we find that our 18-module string produces a maximum V_{oc} of 369 volts, which is rather high.

Here the system designer has to make a judgment call. The string size could be reduced from 18 to 16 series modules. (You could have 17 modules in a series string; however, strings of prime numbers should be avoided because of difficulty in physically configuring the array mounting.) Sixteen series modules would give us a high open-circuit voltage of 328, which gets us within the 338-volt upper limit while pretty well keeping us within the 200- to 300-volt window on nominal operating voltage, with a range of 197.6 to 260.32 volts.

The judgment call comes into being because, with a series string of 18 modules, the high open-circuit voltage is above the maximum and the PCU may not start up under certain conditions. On the other hand, with a series string of 16 modules, the low nominal operating voltage is slightly below the low cutoff point and the PCU may shut down under certain conditions. The designer must review the temperature extremes for the site to see how frequently they occur and make the best estimate as to which set of operating conditions the array is most likely to see over time. Will the array on average see higher operating temperatures or lower? If the answer is higher, a series string should contain 18 modules. If the answer is lower, a string of 16 modules should be specified.

A review of the temperature information for the site shows that the data represents extreme highs and lows rather than average highs and lows. Further study reveals that there are many more hours of operation in the higher ranges than there are in the lower ranges. With this knowledge, the decision is made to configure the array in series strings of 18 modules each.

The PCU should function well under this operating regime. Even if the temperature drops to −5°C (23°F) and the V_{oc} goes up to 369 volts, the PCU will likely start up on that morning anyway, since the array doesn't just instantly jump up to the high open-circuit voltage value. Instead, it starts low and rises as the sun comes up over the array. This gives the PCU a chance to start up and allows its maximum-power tracker to gain control of the array voltage before it goes beyond limit. The only problems that could occur would arise during a midday start-up condition that followed a power outage or some planned system shutdown. During a midday start-up of this type, the open-circuit voltage might go beyond limit. But if this happened, the PCU would just wait until the next morning to resume operation. Of course, the experienced designer will be well versed in the operating characteristics of equipment being specified and will thus be aware of the PCU's response to variations in temperature and other operating conditions.

With the array string size set at 18 modules in series, we next set out to determine the number of series strings required to deliver the 4.5-peak-kilowatt array output. We know that the modules chosen have a peak rating of 30 watts under STC of 1,000 W/m² and a cell temperature of 25°C (77°F). Eighteen modules times 30 watts delivers 540 peak watts from each series string. Thus, 8.33 strings will be required to deliver the nominal 4.5 peak kilowatts (without correction for insolation and temperature influences).

At this point in the design process the system owner should be consulted, because once we have determined the string size that will satisfy the voltage requirements of the load, the sizing of a UI array is somewhat arbitrary. The utility company will supply whatever power the array cannot and the sizing is (at least currently) not driven by the economics of power sale. The client must decide how much money he wants to invest in his array. Any system configuration from one series string of 18 modules up to and including ten series strings will function with the chosen PCU. The optimum system size with regard to PCU efficiency, in this case, would be between eight and ten strings. My recommendation would be nine or ten.

A nine-string array would be rated at 4.86 peak kilowatts under STC and, when corrected for the actual average temperature and insolation conditions that exist at the site, would deliver an average of approximately 3.6 kilowatts. The array would be mounted at an altitude angle approximately equal to site latitude unless special considerations dictated otherwise. For example, if loads are constant year-round, the decision may be made to tilt the array more toward the winter sun. On the other hand, since more excess power can be generated for sale to the utility in the summer, the owner may prefer to tilt the array more toward the summer sun. Another consideration may simply be the pitch of the roof on which the array will be mounted. Thus, the angle equal to site latitude represents a convenient median point for design purposes.

The preceding examples have presented the basic principles involved in the sizing of a photovoltaic array. We have seen that peak power ratings of modules based on the STC criteria of 1000 W/m² insolation and operating cell temperature of 25°C are unrealistically high to be used in design calculations. We have demonstrated the need to establish a nominal operating cell temperature for the chosen module that more accurately reflects the field conditions at a particular site. A NOCT of 45° or 46°C is commonly chosen, though site-specific variables should be taken into account, especially when working with larger systems.

In this chapter we also contrasted two different procedures for sizing photovoltaic arrays: the amp-hour method and the watt-hour method. The first is commonly used when designing stand-alone systems that charge a battery; the second is used when designing SA systems without storage and for utility-interactive systems. We will encounter more examples of array sizing when we focus on design case studies in later chapters.

CHAPTER FOUR
VOLTAGE REGULATORS, BATTERY CHARGERS, AND STORAGE BATTERIES

In chapter 2, we reviewed the basic options available for residential and commercial photovoltaic (PV) installations. Residential options include stand-alone (SA) systems that employ direct current (DC), alternating current (AC), or a combination of the two, and utility-interactive (UI) AC systems. At the top of the scale of commercial installations is the central-station AC generating system. However, in our discussion of hardware, we have up until now concentrated most of our attention upon the single component common to all of these configurations, the PV module itself. The primary emphasis given to modules is understandable, since they constitute the "input" end of a solar electric system. However, in the vast majority of applications, PV modules must be teamed with other hardware to satisfy load requirements.

This is just as true for small residential installations as it is for central-station utility power plants. A huge photovoltaic array field (or, for that matter, a hydro-dam or coal-fired generator) does not, in itself, make a power grid. Neither does a roof-mounted PV array, by itself, constitute a viable home electricity source. Electricity has to be transformed (if alternating current is required), transported, and in some cases stored, before the generating system becomes useful. The devices that serve these purposes in a PV installation are called the *balance-of-system* (BOS) components, and their selection and proper integration into the overall design are essential to the efficient functioning of the PV system.

Balance-of-system components—**the parts of a photovoltaic system supplementary to the modules**

In this and the following chapter, we will take a component-by-component look at those parts that make up the balance of a photovoltaic system. The focus of this chapter will be on voltage regulators, battery chargers, and storage batteries, as well as on the hardware that safely and efficiently gathers power from the modules prior to storage. In chapter 5, we will move on to consider the equipment specifically required by AC systems: DC-to-AC inverters, backup generators, and controls.

Lightning Protection

High-voltage transients—electrical currents in the atmosphere generated as a result of nearby lightning activity

As electricity flows from the array through the system toward the load, the first specialized component it should encounter is a lightning protection device. The threat that lightning activity poses to a conventional house is compounded in the case of a photovoltaic-powered residence by the potential damage to electronic BOS equipment by high-voltage surges. I am not talking about protecting a PV residence or its power system from a direct lightning strike, which in the vast majority of cases is a very unlikely occurrence. The challenge here is to protect the system's sensitive electronics components from the far-more-frequent *high-voltage transients,* which can be caused by lightning activity some distance away from the site.

The traditional "lightning rod" method of direct-strike protection is not recommended. Not only does it not offer the needed protection, it is also quite expensive and may deliberately attract large voltage surges in the direction of the photovoltaic system.

In a properly installed photovoltaic system, some protection is provided by the DC (and AC) circuit and array frame (mechanical) grounds. However, a simple frame ground is often not sufficient to prevent costly damage to system-control electronics and inverter circuitry if a high-voltage surge enters the equipment. In a utility-interactive PV system, a high-voltage surge from a lightning strike on or near the utility distribution feeder can easily find its way into an inverter if the AC lines don't have lightning protection.

The best way to provide reliable, low-cost protection for sensitive electronic equipment is to use the solid-state devices called *surge arrestors.* The surge arrestor most commonly used in photovoltaic systems is the metal oxide varistor (MOV). In addition to the MOV, other transient protectors employed in PV systems include zenor diodes, transorbs, and gas-discharge surge arrestors. (Instructions for installation of surge arrestors can be found in chapter 9.)

Surge arrestor—a device tripped into short-circuit by a sudden surge of high voltage such as that produced by a nearby lightning strike

In a properly designed photovoltaic system, all nongrounded DC and AC conductors will have surge arrestors. Of course, these devices will not protect installations from direct lightning strikes. However, they will prevent high-voltage surges caused by a nearby strike from ruining valuable equipment.

The surge arrestor functions in an open-circuit mode until a high-voltage surge trips it into short circuit. This happens when the surge exceeds the arrestor's breakdown voltage, usually 40 to 60 volts (V) of direct current (VDC) in small 12-volt photovoltaic systems and 320 VDC or more in larger, higher-voltage PV applications. Some good-quality surge arrestors will reopen and continue to function after the danger is past in all but the worst-case situations. When a high-voltage surge occurs with an inexpensive surge arrestor, the arrestor will conduct the surge to ground, protecting the system but destroying itself in the process.

Besides breakdown voltage, other features to look for in a surge arrestor are a sufficient current rating (40 amps or higher is usually recommended) and a maximum energy rating, measured in a unit known as joules (J), of 100 joules or more. A surge arrestor is often installed in a housing of its own, which can be located either outside or inside the building. Surge arrestors are also sometimes installed inside the string combiner box, which I will now describe.

String Combiners

The physical point at which the leads from the photovoltaic array strings or source circuits are joined in parallel to create the main array output leads is called the *string combiner.* It is essentially a sophisticated junction box, the place where all of the positive and all of the negative leads from the strings are brought together after being protected from high-voltage transients by the surge arrestors just described.

The string combiner can be built on the job site by an electrical contractor. However, the inclusion of input current test points, blocking diodes, and current shunts makes its manufacture best left to firms that specialize in power-conditioning hardware, especially if you're working with a large array. Also, in the future this component will probably have to be approved by Underwriters Laboratories. When that becomes the rule, the string combiner will have to be purchased as a complete, premanufactured unit.

PHOTO 4-1

A string combiner for two array strings, manufactured by Ascension Technologies.

The advantages offered by the combiner are considerable. Without it, both series and redundant parallel cross-wiring of strings would be necessary. This would make the photovoltaic array one discrete electrical unit, which means that testing, troubleshooting, and identifying specific problem areas would be very difficult. With the string combiner, a panel of DC input isolation switches and test points can be used to isolate individual strings within the array. This makes I-V curve charting and other diagnostic procedures far more simple and precise. A schematic diagram of a typical string combiner box, which includes surge arrestors and blocking diodes, is shown in figure 9-1.

Blocking and Bypass Diodes

A device called a *blocking diode* also helps to manage the flow of electricity in photovoltaic systems. In an SA system, when the PV array is generating a voltage that is *less than* the battery voltage, such as at night, electricity stored in the batteries can and will flow back out into the array and be dissipated as heat. If allowed to occur, this reverse flow will drain the batteries and could also damage the array.

Blocking diodes allow the current to flow in only one direction (from the array to the batteries) and are included in the photovoltaic array wiring to prevent reverse current flows. They are installed in series with the positive conductor from each array string. The diode's anode is turned toward the PV array and the cathode toward the load. Some voltage regulators are furnished with internal blocking diodes already installed, while other types eliminate the need for them by using a relay to disconnect the array at night. In many stand-alone and nearly all utility-interactive photovoltaic systems, blocking diodes are located in the string combiner box.

Blocking diodes are needed in UI as well as SA photovoltaic systems because they protect the individual series array strings from the possibility of heavy reverse current flow due to ground faults. If a fault to ground occurs in a string and there are no blocking diodes, all the current from the remainder of the array will flow backward to ground through the modules in the affected string. Such high reverse current flow, if unchecked by a blocking diode, will likely cause serious damage to the modules from excessive heat. If the array is very large, it could even cause a fire.

There is one key drawback to blocking diodes. Their use leads to a certain amount of power dissipation, typically 0.75 to 1 watt per amp for a silicon diode and 0.50 watt per amp for a Schottky diode. Schottky diodes, it should be noted, have lower wattage ratings than silicon diodes and are two to three times more expensive. Both types of diodes will dissipate array power and produce a drop in array voltage, which in some system configurations will contribute to the system's internal (parasitic) losses. In many cases, blocking diodes will also require a heat sink mounting.

If one module in a series array string becomes damaged—or is shaded by trees, chimneys, or other objects—so that the module does not produce electricity, it will limit the current flow in all of the other modules in the series string. The effect is much like that in the old series strings of Christmas tree bulbs: when one bulb failed, they all went out. A small but important device called a *bypass diode* provides an alternate path around the failed module for the flow of current produced by the remaining modules

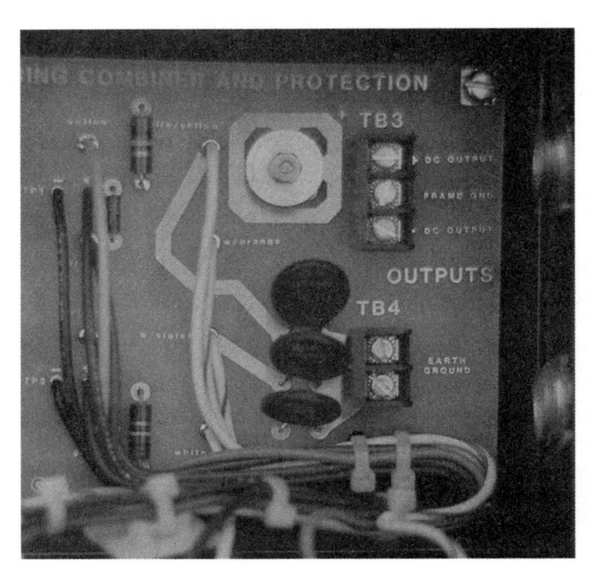

PHOTO 4-2

The interior of a string combiner box, showing three lightning surge arrestors (center) and a blocking diode (top).

in the string. Without the bypass diode, the current produced in the good modules left in the string would be "dropped" across the failed module(s) and dissipated as heat.

Typically, one bypass diode is placed in parallel with each module in a series string, with the diode's anode on the module's negative terminal and the diode's cathode on the module's positive terminal, as shown in figure 4-1.

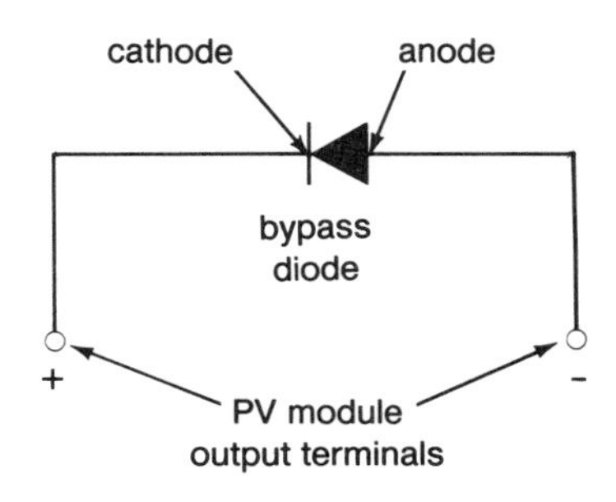

FIGURE 4-1
A diagram of a bypass diode installed across the output terminals of a photovoltaic module.

With a bypass diode in place, the *string current* will remain essentially unchanged when one photovoltaic module in a large series string of modules is shaded; the *string voltage* will decrease by an amount equal to the normal operating voltage rating of the module plus the forward voltage drop of the diode, which is typically around 0.7 volt.

Bypass diodes are important items to include in all but the most basic 12-volt photovoltaic arrays, and many module manufacturers now install them in each module as a standard feature. In early PV systems, which didn't have bypass diodes, a lot of modules were destroyed when an entire string's worth of current was "dropped" across a single cell made inoperative by shading or damage. Such a cell became an electrical resistor rather than a current producer. The resultant heat from the dissipated current created a "hot spot" condition, which over time led to module delamination and failure. Therefore, you should always make certain that each PV module in a series string is protected by its own bypass diode. Also, you should make every effort to install the PV array in a location where it will not be subjected to heavy shadows from fixed objects at any time during the day.

Voltage Regulators

In photovoltaic systems that interface with the utility grid and incorporate no provision for storage, voltage regulation is taken care of by the utility since the UI inverter has no choice but to operate at the utility-supplied voltage. However, when the electrical output of a PV array is stored in batteries, it is necessary to include a separate voltage regulator as an intermediate component between the array and the batteries.

The basic purpose of a voltage regulator is to control the flow of current from the array to the storage batteries in an SA photovoltaic power system and to maintain the state of charge of the batteries, preventing them from overcharging or undercharging. Overcharging is undesirable for a number of reasons. It can cause corrosion and buckling of the lead plate, loss of battery electrolyte, and a buildup of hydrogen gas. All of these phenomena shorten battery life.

Ordinarily, the batteries will be better off if they are not charged at full current right up to the maximum level. Thus, as they approach full storage capacity, the voltage regulator reduces and then terminates the flow of incoming power from the photovoltaic array. While it is possible in some small PV systems to minimize overcharging without a voltage regulator through a combination of careful PV module selection and system design, this practice is dependent on a variety of variable factors and is not generally recommended.

Many voltage regulators also include circuitry that allows them to provide other control functions for the photovoltaic system as a whole. There are, in fact, a lot of different types of voltage regulators. We will now describe a variety of regulator options and the functions they perform.

Voltage regulators differ, in part, according to the manner in which they regulate the flow of electricity into the batteries. The two most common types are the *series regulator* and the *shunt regulator.* Two other types less commonly used are the *sequential* or *array-shedding regulator* and the *ampere-hour integrating regulator.*

The Series Regulator

The series regulator, shown in figure 4-2, incorporates a series switch, which can be either an electromechanical relay or a solid-state transistor and is placed between the photovoltaic array and the batteries. The internal details and operation of series regulators vary from manufacturer to manufacturer and even from model to model within one manufacturer's line. As an example, let's look at the method of operation of the SCI Charger 1, a regulator manufactured by Specialty Concepts Incorporated.

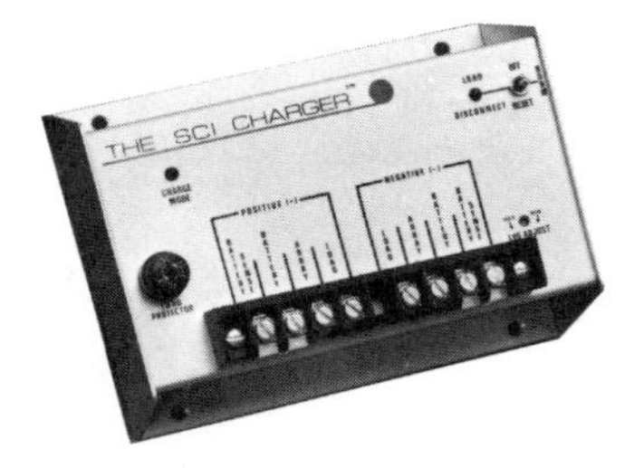

PHOTO 4-3
The SCI Charger series-type voltage regulator manufactured by Specialty Concepts.

In the morning as the sun rises, the regulator senses the rise in array voltage and closes the charging relay, thereby connecting the array directly to the batteries. (The relay closing process draws approximately 0.14 amps out of the power system.) This allows current to flow from the array into the batteries and also to the load as required. The SCI Charger 1, along with certain other series regulators, features a two-step charging sequence. In units of this type, when the battery voltage increases to a preset level called the *high-voltage level termination point,* the series relay opens.

The high-voltage termination point is indicative of a 90 to 95 percent state of charge and is set at approximately 14.5 to 15 volts on a 12-volt system, depending on manufacturer and system parameters. When the relay is open, the full array output is unable to flow into the battery. In this way, overcharging is prevented.

Once the high-voltage termination point is reached, the series regulator with two-step charge control goes into the "float charge" mode. In this mode a bypass transistor in parallel with the charging relay contacts allows up to 3 amps of array current to go into the battery. The battery will be maintained at a high state of charge as long as the 3-amp float charge output exceeds the load. When the load needs more energy than the array can deliver through the regulator in the float charge mode, the battery must supply the extra. This will cause the battery voltage to drop. When the voltage falls to the *full charge resumption point,* the charging relay will close, once again allowing full array output to go into the battery while also servicing the load.

At night, the charging relay is opened by a timer that is activated 10 hours after sunrise. If it finds the array still producing current, the relay closes and reopens 2 hours later. This process is repeated every 2 hours

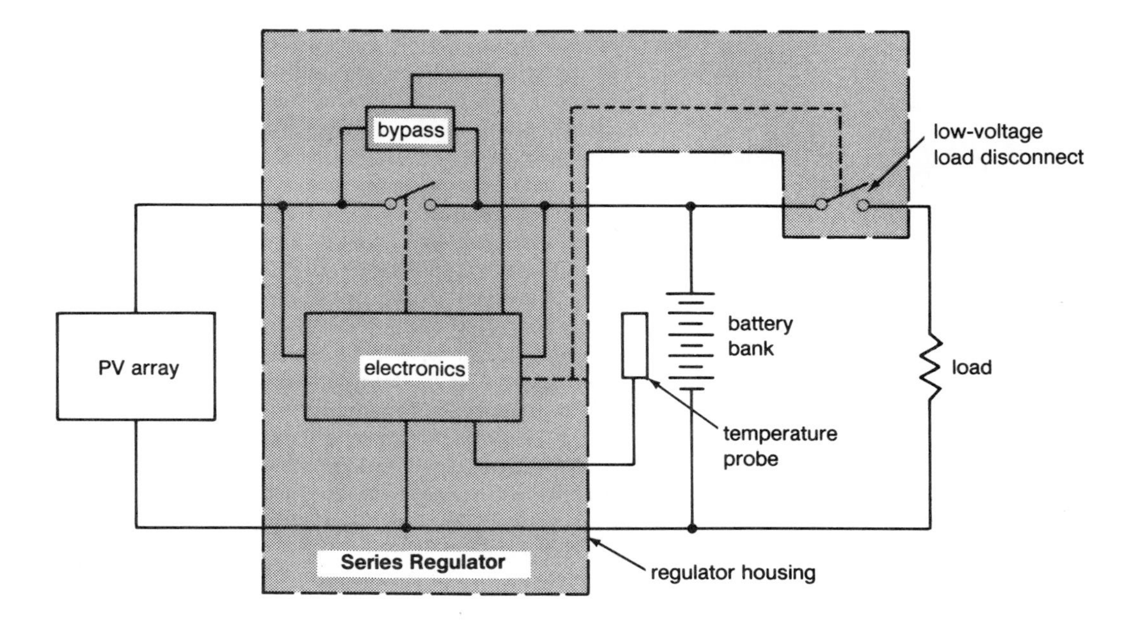

FIGURE 4-2

A block diagram of a series-type voltage regulator.

until it finds the array has ceased producing current. At that point, the relay stays open until sunrise.

The charging relay in this regulator is made in such a way that a small amount of energy is needed to keep it closed. Leaving the relay open during the night conserves that energy. More important, it eliminates the need for a blocking diode to prevent the batteries from discharging themselves through the array. On days with intermittent cloudiness, short cycling of the charging relay is prevented through the use of a time delay and the choice of a threshold voltage that is below the voltage of the photovoltaic array for "typical" cloudy periods.

One advantage offered by series regulators is that they can be purchased in a wide range of current-handling capacities and require no blocking diode. Disadvantages include the parasitic loss due to the relay's power consumption and the fact that less expensive models use electromechanical switching, which is less reliable than solid-state switching.

The Shunt Regulator

The shunt regulator shown in figure 4-3 is a solid-state voltage regulator that prevents overcharging of the battery through the use of a transistor that shorts or bypasses the output of the array. As with series regulators, internal details and operation of shunt regulators vary from model to model. The regulator described here is the Solar Sentry model made by Balance of Systems Specialists (BOSS), Inc.

When the sun rises, so does the voltage of the photovoltaic array. When the array's voltage becomes higher than that of the battery bank, battery charging commences. The regulator delivers the array's full charge current to the battery until its control circuitry senses a high battery voltage (the previously mentioned high-voltage level termination point). At this point, the regulator's transistor goes into the shunt mode, shorting the array and dissipating its electrical output within the array as heat.

While in the shunt mode, the regulator also functions in a pulsating, "charge-pumping" manner; that is, the shunt transistor opens and closes at varying rates—every 30 seconds on the average—depending on the array current. This ensures that the batteries receive a full charge and remain charged. If during this cycling the regulator discovers that the batteries have been discharged and their voltage is now below the high-voltage termination point, it will switch out of the shunt mode and resume full charging.

PHOTO 4-4
The Solar Sentry shunt-type voltage regulator manufactured by Balance of Systems Specialists.

A shunt regulator incorporates a blocking diode to keep the batteries from discharging through the array at night. Because it dissipates power, a blocking diode can become quite hot and often requires a heat sink. The shunt transistor also must be mounted on a heat sink. A series regulator, by contrast, needs no blocking diode since it relies on the charging relay to disconnect the array. Because of the heat sinks, shunt regulators are usually bulkier than series regulators, especially those models rated to carry 10 amps or more.

There are two major problems with shunt regulators. First, the majority of them cannot tolerate the array being connected while the battery is disconnected. Also, most models cannot tolerate an accidental reversal of battery polarity during installation. Either condition usually means certain death for the regulator. Fortunately, the newest units on the market, such as the BOSS Solar Sentry, are not plagued by these shortcomings.

For 12- and 24-volt applications, shunt regulators typically have a current rating between 6 and 20 amps, whereas series regulators commonly

FIGURE 4-3
A block diagram of a shunt-type voltage regulator.

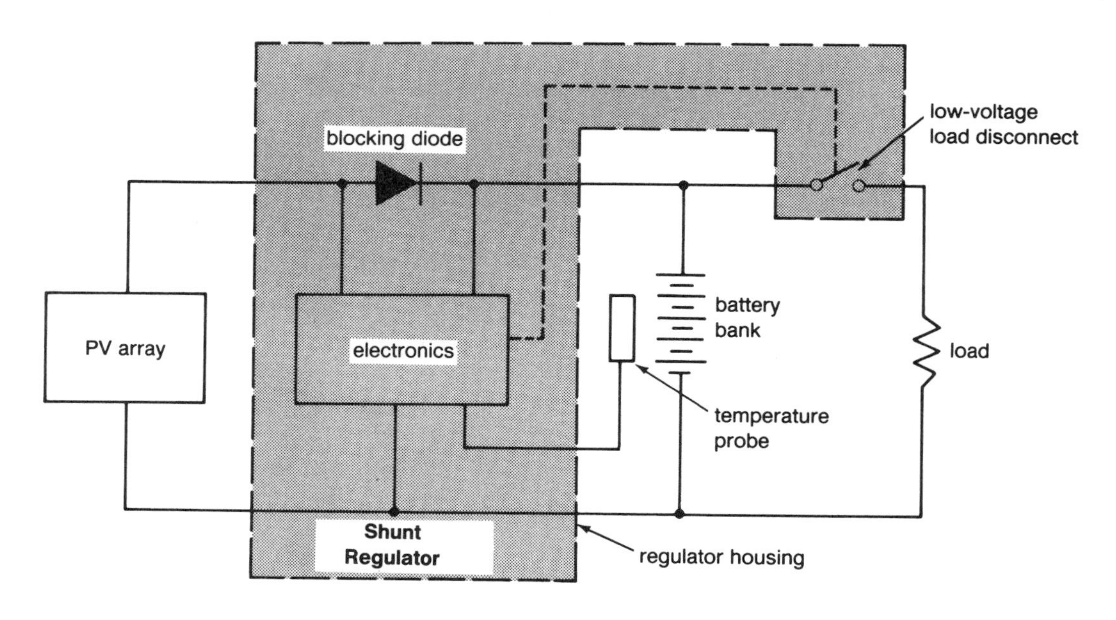

have ratings up to 30 amps. In systems that produce greater current than a single regulator will carry, several regulators may be used if they are connected in parallel. Regulators of either the series or shunt type may be connected in this manner, although you should always check with the manufacturer regarding individual unit capabilities and only connect together regulators of the same make and model.

An innovative method used by Balance of Systems Specialists for increasing current-handling capability of individual regulators involves the addition of "slave" units—that is, separate switches lacking internal control mechanisms. In their Master-Slave Series, BOSS employs a master control board to direct individual slave units. Each slave is capable of handling the same amount of current as the master unit, typically 12 amps in a 12-volt system. Master-slave systems are available in voltages ranging from 12 to 120 with total current ratings up to 600 amps.

Another clever BOSS innovation is the design of a regulator that will charge two separate battery banks of different voltages. The BOSS Tandem Series allows you to charge both a 12-volt and a 24-volt battery bank from the same array. It also features priority charging and load management switching that allows you to charge one battery bank before the other and provides the means for disconnecting the load before either battery bank is excessively discharged.

The Sequential Regulator

The sequential (array-shedding) regulator is used on large, multi-kilowatt photovoltaic arrays. The large array is divided into subarrays, each of which is controlled independently. As the state of charge in each battery bank reaches its maximum level, the subarray feeding it is disconnected. The reverse is also true: the subarrays come back onto the system as the state of charge in their respective battery banks declines. This regulator is typically made up of multiple series-type regulators and is custom-configured for each individual application.

One advantage of the sequential regulator is its tolerance of failure. If one of the series regulators that make up the system fails, only a portion of the total array charging current is lost.

The Ampere-Hour Integrating Regulator

The amp-hour integrating regulator provides the most accurate means of controlling the state of charge of a battery bank because it measures and logs the number of amp-hours entering and leaving the battery bank. By contrast, voltage-sensing regulators (series, shunt, and sequential) measure only the two battery charge/discharge limits. The state of charge of a battery bank can vary significantly for a given voltage, depending upon the charging or discharging current. (This is especially true for an old battery, which is more erratic in performance than a new one.) The amp-hour integrator can also more accurately limit the battery depth of discharge.

The amp-hour integrating regulator is primarily suited to high-current systems because its parasitic losses are greater than those of series or shunt regulators. Due to the added electronic components and memory storage inherent in this type of regulator, it is perhaps somewhat less reliable than other types. These units are not currently available "off the shelf" and must be custom-built for a particular application. However, I expect that this situation will change as the market for photovoltaic system electronics increases and the many advantages of the amp-hour integrating regulator become more widely recognized.

Optional Features for Voltage Regulators

A variety of features may be incorporated into each of the voltage regulators previously described. We will briefly describe some of the more common ones.

A *low-voltage load disconnect* cuts off the load at a preset battery voltage level called the *load disconnect threshold* (11 to 12 volts for a 12-volt system). This protects the battery bank from being discharged too deeply. The load is reconnected after the battery voltage rises to a higher level, called the *load reconnect threshold* (between 13 and 13.6 volts for a 12-volt system). The gap between these set points prevents rapid on/off cycling of the load.

Adjustability is desirable in a regulator's set points. This is true for both the points at which the load disconnect is engaged and disengaged and the points at which battery charging starts and stops. Many of the higher-quality regulators on the market now include this feature. However, some units are difficult to adjust accurately in the field.

Regulators with *battery temperature compensation* automatically adjust the high-voltage charge termination point to maximize the charge entering the battery bank with respect to the battery's temperature. The compensation made for temperature excursions from a base of 25°C (77°F) is on the order of 0.005 volt per degree Celsius of temperature rise per battery cell for lead-acid batteries. If the temperature-sensing device is located inside the regulator, the regulator must then be installed very close to the battery bank. However, the regulator can be furnished with a remote battery temperature probe that allows it to be placed outside of and away from the battery enclosure. Such exterior mounting is preferable to avoid the ignition of any hydrogen gas that may be caused by the electrical arcing produced at the relay contacts during the regulator's operation.

Two common *instrumentation* options found on voltage regulators are DC current and voltage meters. Other parameters that can be measured include amp-hours (Ah), watts (W), watt-hours (Wh), and battery state of charge. Typically, these tasks require custom measurement equipment configured to the particular application. I recommend that DC voltage and current meters be included in all but the smallest PV systems. Often this is most easily done by adding them to the voltage regulator.

Alarm circuitry is often used to signify a condition of unusually high or low battery voltage. When this condition occurs, a switch is turned on that activates an alarm signal. The signal may be a light, an audible alarm, or a radio wave. Alarms are frequently used in remote systems of critical importance. Alarm or status signals can also be used to indicate the following conditions: load disconnect, battery charging, battery fully charged, low battery charge, battery polarity incorrect, array polarity incorrect.

Transient protection is often a standard feature on quality regulators. It is very important because high-voltage surges can be very destructive to the electronics in the voltage regulator and other systems components. A high-voltage surge can result from a nearby lightning strike, as previously described, or from a load that causes a rapid on/off current flow. Rapidly oscillating current flow can be caused by DC motor brushes, DC contactors, or by DC-to-AC inverters when the switching frequency of the inverter interferes with the electronic logic of the voltage regulator.

Selecting a Voltage Regulator

There are many factors to consider when choosing a voltage regulator. The type of photovoltaic system with which you are working will determine which are most important, so let's take a brief look at each one.

Obviously, *reliability* is a desirable feature in all components that make up a solar electric system. However, some systems need to be more reliable than others. For example, it is more critical for communication repeaters to be totally reliable in their operation than it is for street lights or residential lighting. The degree of reliability built into a given system is thus a function of the system's importance to the user and the amount of money he is willing to pay for it.

Power-handling capability is predetermined by the array and load size. It is always important to choose a regulator that can match or surpass the power-handling requirements of the array and load.

A *low-voltage load disconnect* is a very useful feature because it accomplishes two things. First, it extends battery life by not allowing a battery to drop below its rated capacity. Second, it decreases the likelihood of damage to the load in cases where the load has a minimum input voltage requirement.

The method by which the regulator isolates the array at night is an important factor in determining *parasitic loss,* which in small systems (100 watts or less) may be a considerable portion of the total system's design load. In order to avoid having to increase the array/battery size, it is best to select a regulator that causes a negligible parasitic loss.

Completely *solid-state componentry* is sometimes necessary where there is danger of explosion, as in a gaseous environment of a poorly vented battery enclosure. Also, a corrosive atmosphere of the sort found in coastal areas or offshore oil platforms can cause mechanical relay contacts or circuit board contacts to degrade. In situations of this type, only voltage regulators that have all solid-state componentry should be used.

The *size* of the regulator is important only with respect to its intended location in the photovoltaic system, such as when it must be placed in the battery enclosure, and how far it must be transported to the installation site. A very large regulator can be very expensive to transport to a remote site and can entail high labor costs for installation.

Ease of installation is a very practical consideration. The easier it is to connect a regulator into a system, the less time and money you will spend. This is especially true if periodic replacement of the regulator is necessary. Positive things to look for are a terminal block that will accept crimp-on lugs large enough for the required conductors, good installation instructions, and solid construction.

An investment in *battery temperature compensation* circuitry in your voltage regulator will pay for itself many times over in enhanced system performance and extended battery life. Battery temperature compensation should be included in all but the most basic photovoltaic systems unless the batteries will be kept within a temperature range of 12° to 32°C (54° to 90°F).

Table 4-1. Comparison of Various Voltage Regulators

Manufacturer and Model	System Voltage (V)	Array Current (Max. A)	Load Current (Max. Resistive/A)	Types of Control
BOSS Solar Sentry	12, 24	10, 20	15	Switching shunt with Schottky diode
Specialty Concepts SCI Charger	12, 24	30, 20	30, 20	Series relay without blocking diode
TRI Solar BCR	12, 24, 36, 48	25	15 @ 24 V max.	Shunt with blocking diode
BOSS Power Control	Model 1—12, 24 Model 2—48, 120	Model 1—30 Model 2—20	Model 1—30 Model 2—20	Series switching (Optional Schottky diode)
BOSS Econocharger	12	5	(Optional)	Switching shunt with Schottky diode
Specialty Concepts ASC	12, 24	2, 4	N/A	Shunt with blocking diode

Warranty and service are important considerations. Find out the terms of the complete warranty and what is required to obtain service or replacement should the regulator malfunction.

The *cost* of a regulator is dependent on the size of the array and load. Other factors that affect cost are the regulator type, the method of metering employed, and the quality and type of enclosures used (rainproof enclosures, for example, are expensive and often necessitate the addition of a heat sink).

Given these parameters, experience has shown that quality units of both the series relay and shunt type have given the most reliable performance in small to medium-size residential photovoltaic systems. (Comparative data on six different voltage regulators is provided in table 4-1.)

Low-Voltage Load Disconnect Point (V)	Low-Voltage Load Reconnect Point (V)	High-Voltage Charge Cutoff (V)	Trickle Charge Current (Max. A)	Temperature Compensation
(Optional) Adjustable 11.5, 23.0	13.5, 27.0	(Optional with lead-antimony battery) Adjustable 14.475, 28.950	N/A	(Optional) Remote/local integrated circuit
11, 22	13, 26	14.1, 28.2	3, 1	Remote resistive temperature device
(Optional) 10.7, 21.4, N/A, 42.8	11.7, 23.4, N/A, 46.8	14.4, 28.8, 43.2, 57.6	25	Remote thermistor
Model 1—adjustable 11.5, 23.0 Model 2—adjustable 46, 115	Model 1—adjustable 13.5, 27.0 Model 2—adjustable 54, 135	Model 1—adjustable 14.475, 28.950 Model 2—adjustable 57.90, 144.75	N/A	Remote integrated circuit
Adjustable 11.5	13.5	Adjustable 14.475	N/A	Local integrated circuit
(Optional) 11.52, 23.04	(Optional) 13.02, 26.04	14.1, 28.2	2, 4	N/A

continued

Table 4-1. Comparison of Various Voltage Regulators—*continued*

Manufacturer and Model	Quiescent Current Drain (mA)	Worst-Case Current Drain (mA)	Indicator LED's	Meters
BOSS Solar Sentry	15–30	110, 75	Charging; battery discharged; array & battery polarity; low battery	N/A
Specialty Concepts SCI Charger	10	180	Low-voltage disconnect	Battery voltage; array current
TRI Solar BCR	8, 6, 5, 6, 6	7.5–65.0	N/A	N/A
BOSS Power Control	15–30	Model 1—170, 115 Model 2—75, 58	Charging; charged	Array current; load current; battery voltage; adjustable set point (Optional digital display)
BOSS Econocharger	15–30	110	Charging; charged (Optional polarity protection)	N/A
Specialty Concepts ASC	5, 7	5, 7	N/A	N/A

Batteries

There are two basic types of batteries: the *primary battery* and the *secondary battery.* A primary battery is given an initial charge during its manufacture. This charge wears down with use, after which the battery is discarded. The most familiar examples of this type are flashlight batteries, the larger lantern batteries, and the little button-size power units used in watches and calculators. In contrast to primary batteries, secondary batteries can be repeatedly discharged and recharged, a process known as *cycling.* Not so long ago, primary batteries were used alone to satisfy many

Fuses	Operating Temperature (°C/°F)	Enclosure	Size (in)	Weight (lb)	Warranty
N/A	−20 to 50/ −4 to 122	Metal, epoxy heat sink	10A—8 × 4 × 1.3 20A—8 × 4 × 2.5	10A—1.2 20A—1.8	1 year
Standard (Optional circuit breaker)	0 to 50/ 32 to 122	Vented box	8 × 5.5 × 3	2.0	1 year
N/A	−25 to 60/ −13 to 140	Weatherproof box	6 × 9 × 3.5	2.5	1 year
Resettable battery circuit breaker; array & load switch	−20 to 50/ −4 to 122	NEMA 3R	Model 1—10 × 8 × 4 Model 2—12 × 12 × 4	8.0	2 years
N/A	−20 to 50/ −4 to 122	Metal, epoxy heat sink	4 × 4.5 × 1.5	0.8	1 year
N/A	−20 to 50/ −4 to 122	Encapsulated	4.7 × 2 × 0.9	0.25	1 year

remote power requirements such as offshore navigational aid systems. Now, more and more, these primary batteries are being replaced by photovoltaics and secondary batteries.

Secondary batteries used for storage in photovoltaic systems must be capable of accepting a large number of charge/discharge cycles. But not all secondary batteries are alike, since they are manufactured from a variety of materials and by a number of different methods. The more commonly available types are *lead-acid* and *alkaline* (including *nickel-cadmium*). In the past, the high cost of nickel-cadmium batteries (also known as nicads) precluded their use in PV systems. Today, in locations where extreme weather conditions prevail (very cold and/or very hot), nicads may compete favorably in terms of life-cycle costs.

Other battery technologies—including zinc oxide, sodium sulfur, and lithium sulfur—are under development. However, for the foreseeable future, the most realistic option for home photovoltaic systems will remain the lead-acid battery, in either its lead-calcium or lead-antimony form. One important feature that distinguishes various models from one another is the depth of discharage for which each was designed.

Shallow-cycle batteries can be discharged to between 10 and 20 percent of their capacity in ampere-hours on a daily basis. For example, a 100-amp-hour shallow-cycle battery will deliver 10 to 20 amp-hours on a daily basis (provided it's also recharged daily). Repeated deep discharges, on the order of 50 to 80 percent of rated capacity, will very seriously shorten the life of these batteries. With proper design, however, this type of battery can serve quite well in a photovoltaic system.

Deep-cycle batteries, by contrast, are designed to allow a daily discharge of between 60 and 80 percent. They are typically more expensive than shallow-cycle batteries; however, fewer units are required to deliver the equivalent usable storage capacity. The deep-cycle battery is most often the battery of choice for residential photovoltaic power systems.

Battery Construction and Operation

A lead-acid battery is typically constructed of lead or lead-alloy plates immersed in a sulfuric acid/water solution called the *electrolyte.* These plates are either positively or negatively charged. Each *cell* of the battery consists of a group of positive and negative plates with a single positive and a single negative output terminal. The nominal voltage of a cell is 2 volts. A 12-volt battery, then, has six cells that are connected in series. A battery's cost, operative life, and efficiency depend on the quality of its plate materials and construction.

Lead-acid batteries store and release electricity by a process known as an *electrochemical reaction,* which involves a series of chemical changes and a flow of electrons. This reaction takes place between plates of lead alloy and the sulfuric acid/water electrolyte solution. The positive plates in the battery are composed of porous lead dioxide (PbO_2) paste supported on a lead mesh grid, while the negative plates consist of sponge lead (Pb). (For simplicity, I am eliminating the chemical symbols of the additional elements—usually calcium or antimony—that combine with lead to form the alloys used in battery plates.) The plates are separated within the battery case by panels made of microporous rubber or polyvinyl chloride (PVC) plastic, and are surrounded by the electrolytic solution.

When an incoming charge current is applied to the battery, the reaction begins: the porous lead dioxide in the positive plates is transformed into lead peroxide, while the lead in the negative plates increases in sponginess. The negative cycle, in which the battery gives off current, constitutes a reversal of this process, with the spongy lead in the negative plates becoming lead sulfate. The electrolyte also changes during the charge/recharge process. The volume of water relative to sulfuric acid in the solution decreases as the battery is charged and increases during the discharge cycle.

Lead-acid batteries are constructed with several different types of plate and grid composition, depending on the application. The *pure lead* (lead-lead oxide) battery typically has a very long life (25 years or more), with high current-rate discharge capability. Its positive plates contain about 2 percent antimony. This battery has historically been used in standby power applications; however, it is beginning to see service in some photovoltaic systems.

Lead-antimony batteries have a concentration of 2.5 to 4 percent antimony in their positive plates. They are designed to tolerate a deep discharge and have very good charge/discharge cycling capabilities. They also have higher self-discharge rates than most other battery types and this progressively increases with age. However, this phenomenon will have little or no impact on a battery storage system that is cycled daily.

Another drawback to this type of battery is that a good deal of water vapor is produced by gassing of the electrolyte during the charging process. If the battery undergoes heavy use (extensive generator charging), the water lost in this way must be replenished as frequently as every three months, which increases the system maintenance requirements. The use of catalytic recombining vent caps, such as those made by Hydro Cap,which recombine hydrogen and oxygen and return the water vapor to the electrolyte, will dramatically reduce this maintenance requirement. An added benefit of the Hydro Cap is that it lowers the amount of hydrogen gas released and the amount of sulfuric acid corrosion that occurs on wires and terminals.

The *lead-calcium* battery is characterized by a low self-discharge rate—1 to 4 percent per month at 25°C—and low gassing during charging. Like the lead-antimony battery, it has good cycling capability but is limited in the number of deep discharges it can withstand. It is used only in shallow-discharge applications. The lead-calcium battery requires lower

PHOTO 4-5

A bank of deep-cycle lead-antimony batteries manufactured by Surrette Corporation.

maintenance and in some cases has been designed as a sealed, "maintenance-free" battery (never needing the addition of water).

The *lead-antimony/calcium* battery combines the favorable characteristics of both lead-antimony and lead-calcium batteries. It has a long cycle life (up to 3,500 charge/discharge cycles and more) and allows deep discharge without requiring as much maintenance as the comparable lead-antimony battery. It is also one of the most expensive batteries on the market.

A relatively new innovation in lead-acid battery construction involves the use of a *captive electrolyte.* There are two basic types of captive-electrolyte batteries currently manufactured: the Gelcell battery from the Globe Battery division of Johnson Controls and the Absolyte absorbed-electrolyte battery from the industrial battery division of GNB Battery Company. The basic principle involves the use of a nonliquid electrolyte to replace the liquid sulfuric acid/water solution common to most lead-acid batteries. The Gelcell has a small percentage of silicon dioxide added to the electrolyte, which gives it a jellylike consistency. The Absolyte battery uses a blotterlike material of spun glass between the plates to create a microporous matrix to absorb the electrolyte solution.

Although the basic electrochemistry involved in captive-electrolyte batteries remains the same as that found in ordinary lead-acid batteries, there are many advantages to the new type. The use of a gelled or absorbed electrolyte allows the batteries to be fully charged and sealed at the factory. As sealed units, the batteries are designed to be maintenance free. They have a very low self-discharge rate. This, together with the fact that they cannot leak or spill, makes them much easier to handle, ship, and store.

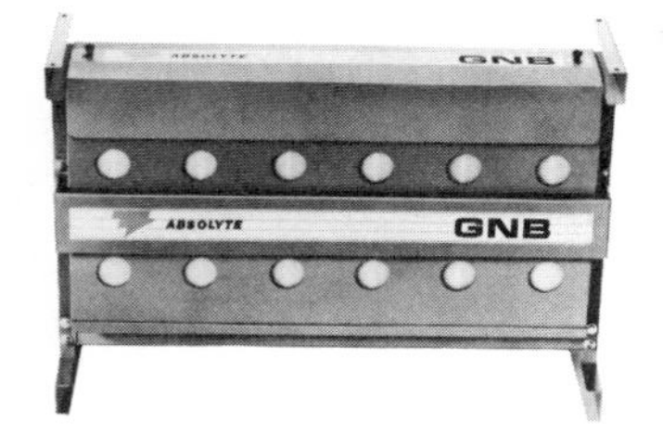

PHOTO 4-6

An absorbed-electrolyte lead-acid battery manufactured by GNB Batteries.

Since the electrolyte is immobilized, as indicated by the term *captive,* electrolyte stratification under heavy discharge is avoided in this type of battery. Captive-electrolyte batteries thus require no equalization charging. They can tolerate lower temperatures and even withstand occasional freezing conditions (however, this is not recommended as a conscious part of the system design). They are available for both shallow-cycle and deep-cycle service, can survive well under extended partial state-of-charge operation, have long cycle lives (over 8 years at 50 percent depth of discharge and over 10 years at 25 percent depth of discharge for some Absolyte models), and can accept higher charging rates than can many liquid-electrolyte, lead-acid batteries.

The captive-electrolyte concept provides a solution to many of the basic problems inherent in lead-acid batteries. And although these batteries are more expensive per unit of capacity than their liquid-electrolyte counterparts, their advantages often outweigh the cost differential. They should be considered for all photovoltaic applications, especially those where site access for regular periodic maintenance is impractical.

Factors Affecting Battery Performance

All batteries work by storing electrical energy in the form of chemical energy. Charging a battery means putting electricity into it, which causes a chemical reaction to take place. When a load is connected to a battery, this chemical reaction goes in the opposite direction and electricity is delivered by the battery.

There is no known energy conversion process that is 100 percent efficient, that has no inherent internal losses. Storage batteries are certainly not exempt from this rule. The overall energy conversion efficiency (or cycling efficiency) of most lead-acid batteries in good condition ranges between 80 and 85 percent. Many factors affect battery performance, and their influence on storage capacity varies from battery to battery. Some are internal to the design and condition of the battery, while others are external. We will briefly look at some of the more important factors.

Battery efficiency—the ratio between the electrical energy flowing into a battery and the electrical energy delivered by the battery to a load

The *daily depth of discharge* (DDOD) is the extent to which a battery is drawn down or discharged by the daily load requirements. This is calculated in terms of the percentage of the battery's full rated capacity. The actual depth of discharge (DOD) will have a strong influence on the storage capacity and life of a battery. If manufacturer's guidelines are not followed and the battery's recommended DOD is exceeded, significant loss of storage capacity and service life can result. For example, a Delco 2000 discharged 10 percent per day will last for nearly five years; however, if it is discharged 20 percent per day it will last only a little more than two years.

The *maximum allowable depth of discharge* (MDOD) is a number provided by the manufacturer that states the allowable extent of discharge in terms of the percent of a battery's full rated capacity. Unlike DDOD, which is a daily routine, MDOD is an infrequent, worst-case design occurrence. It happens occasionally at the end of a long no-sun period when load requirements for several successive days have had to be satisfied directly from the battery without recharging from the solar array. Both values for a given battery vary with respect to the ambient temperature of the battery enclosure.

It is very important to accurately determine the longest likely no-sun period for the site under consideration. If the manufacturer's guidelines are not followed and the battery's recommended MDOD is exceeded in value or reached more frequently than intended, significant loss of storage capacity and service life can result.

The *charging procedure* that is followed can have important effects on battery performance. If a battery is maintained at its float voltage (for the Delco 2000, this is 13.8 volts at 25°C) and never charged above this voltage, two problems can occur. First, the electrolyte may stratify. This means that the acid will be concentrated in the bottom of the battery, with the water at the top. Stratification of the electrolyte results in poor performance and makes the battery more susceptible to freezing and permanent internal damage. Second, in a system where several batteries are connected in series, it is possible that a single battery may lose its capacity before the rest. This will cause poor performance of the entire battery bank.

In order to minimize these problems, it is a good idea to provide an occasional *equalizing charge* to the batteries. This means raising the battery voltage above the float voltage. Under normal charging conditions, the charging rate (the current going into the battery) is lowered once the float voltage is reached so that excess gassing or bubbling of the electrolyte does not occur. This last step in the process is frequently called a *finishing* or *trickle charge*. When an equalizing charge is applied, a controlled amount of bubbling of the electrolyte is purposely introduced to destratify the solution and also bring any weak cells up to full potential. This process is recommended about once a month, but the need for it varies with how hard the batteries are used. Photovoltaic charge control regulators with two-step charge cycles will accomplish this if the PV array is sized large enough in relation to the load and charging requirements. A generator and a battery charger can also be used.

There is an optimal rate at which the chemical conversion of energy that produces electricity in a lead-acid storage battery can occur. (The rate of current flowing into or out of a battery is expressed in terms of amperes per hour.) Most batteries are designed to allow a reasonable latitude in the charge and discharge rates. If these design rates are exceeded, the energy conversion efficiency will be adversely affected and the battery's internal losses will increase. This will result in less net electrical energy delivered.

A battery will deliver its maximum output when discharged at or below the manufacturer's recommended discharge rate. The converse is also true: a battery will accept its fullest charge when recharged at or below the manufacturer's recommended recharge rate. The graph in figure 4-4 presents the manufacturer's data on amp-hour capacity as a function of both temperature and rate of discharge for the Delco 2000.

All batteries continually lose some of their stored energy because of internal reactions, even if they are not being discharged by the load. This is called *self-discharge* and is usually expressed in terms of amp-hours or percentage of capacity lost per month. It increases with battery temperature and age.

In the extreme case of a lead-antimony battery near the end of its life and situated in a hot climate where it does standby service for a photovoltaic array, the monthly self-discharge rate might go as high as 40 to 50 percent of capacity. This means the PV array not only has to supply the battery with the normal charging current, it also has to supply an equal

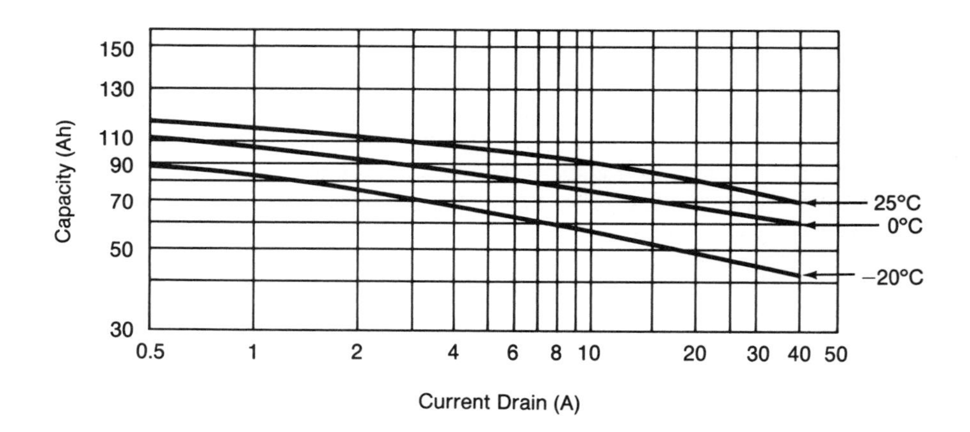

FIGURE 4-4

A graph showing amp-hour capacity as a function of temperature and discharge rate for the Delco 2000 battery.

amount of additional current just to keep the battery from self-discharging. Unless the array is sized to deliver this extra energy, the net effect of the battery's self-discharge on the system is a reduction in the energy available for the load.

In most photovoltaic systems, the battery bank is cycled on a daily basis. When this is the case, the storage loss from self-discharge is almost negligible. Even in a case where there would be a 50 percent loss per month during standby service, the maximum daily loss would be 1.5 percent, which is barely measurable and quickly replenished during the daily cycle.

The *time spent discharged* also affects a battery. A discharged battery should be recharged as soon as possible. Prolonged periods of low state of charge result in severe shortening of its life. If a battery is left in a low state of charge for a long period of time, it either will not regain full capacity or it may not be able to be recharged at all. The time factor will, of course, vary with the type of battery.

A battery's *temperature of operation* is likewise important. The standard battery temperature rating is 25°C. Although a battery operates more efficiently as its temperature increases, it should ideally be kept at or near this point, because as the temperature rises above 25°C, the battery's service life is shortened. In any case, a battery should not be subjected to very low operating temperatures. This causes inefficient operation and can possibly freeze the electrolyte. Freezing becomes even more likely when the battery is subjected to low temperatures while in a discharged state, because the proportion of water to acid in the electrolyte is greater.

The Delco 2000 has a 105-amp-hour capacity (at a 1-amp-per-hour current drain) at 25°C, but only a 95-amp-hour capacity at 0°C (32°F). It is recommended that the Delco 2000 be operated between −5° and 35°C (23° and 95°F). The relative loss of capacity of a shallow-discharge Exide battery is somewhat less: at 0°C it maintains 92 percent of its rated capacity. The graph in figure 4-5 shows the correlation between freezing point and battery state of charge. Note, for example, that a battery will not freeze at −20°C (−4°F) if it is above 50 percent state of charge. However, if it is at a 20 percent state of charge and its temperature is −10°C (14°F), you can count on having to buy a replacement.

Age and *history* take their toll on batteries as well as on most other things. Throughout their service lives, all batteries gradually lose their abil-

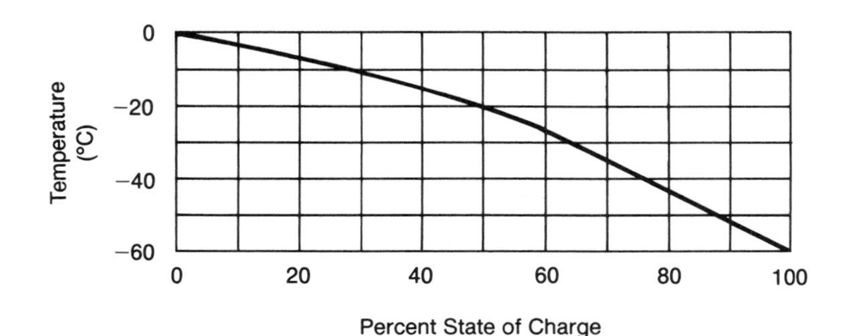

FIGURE 4-5
A graph of freezing point of electrolyte versus battery state of charge.

Useful service life—for batteries this is figured as the length of time in which storage capacity in amp-hours can be maintained at 80 percent of its initial rated capacity

ity to store energy. A battery's useful service life is considered over when its storage capacity in amp-hours is reduced to 80 percent of its initial rating. As a result, in order to ensure that a new battery will continue to satisfy a system's specific load requirement over its full useful life, its nominal rated capacity should be divided by 0.80 to account for the expected loss in capacity. For example, if a 425-amp-hour battery is required, then the nominal capacity of the battery to be specified should be 425 amp-hours divided by 0.80, or approximately 530 amp-hours.

If used batteries are employed, do not expect their performance to equal that of similar but new batteries. Just a few deep discharges on a shallow-discharge battery can significantly shorten its life and capacity. And do not expect to recondition an old battery with some magic formula. Addition of anything other than pure distilled water to a battery will cause problems (short life, low capacity) in the future. Battery additives that promise to improve battery performance or increase capacity should simply be avoided. In general, employing used batteries is not recommended unless past operation and maintenance are well known and the application is considered noncritical.

The *specific gravity* of a battery cell's electrolyte is an indicator of the state of charge of the cell. During cell operation, the electrolyte changes in composition. At full charge the specific gravity is high (1.250 to 1.300) and the acid concentration is high (40 percent at a specific gravity of 1.300). As the state of charge falls, so does the concentration of the acid; acid gets used up and water is formed. Most lead-acid batteries are considered to be completely discharged if the specific gravity is 1.100.

Specific gravity of a liquid electrolyte is measured with a battery cell hydrometer. This instrument should be cleaned with pure water after each use and should not be used to test different types of batteries—that is, don't use the same hydrometer to test both a lead-antimony battery and a lead-calcium battery. The actual specific gravity of the electrolyte in a new, fully charged battery is a part of the battery design and varies with battery type, use, and manufacturer. Although specific gravity of "maintenance-free" cells cannot be measured, some manufacturers have built a state-of-charge indicator into the battery.

Specific gravity readings are affected by temperature. As a result, in order to obtain an accurate value, you must measure the temperature of the electrolyte, record the specific gravity reading, and then adjust it to the corrected value. The adjustment most often used is to add 0.001 to the reading for every 3 degrees above 77°F and subtract 0.001 for every 3 degrees below 77°F. (The comparable process in Celsius would be to add or subtract 0.001 for every 1.66°C above or below 25°C.)

Let's take an example. Let's say the temperature of a particular battery cell's electrolyte is found to be 88°F and the specific gravity reading on the hydrometer is 1.240. We want to determine the corrected specific gravity. Here's our formula:

$$\text{Corrected reading} = 1.240 + \left(\frac{88 - 77}{3} \times 0.001 \right) = 1.244$$

Selecting a Battery

The choice of battery for any photovoltaic system depends on many factors. Generally, deep-discharge batteries are used when fewer than five days of storage are required, while shallow-discharge batteries are more often used when five or more days of storage are required. The reason for this is that in the first case the battery bank is sized to allow the batteries to discharge fairly deeply on a daily basis. By contrast, a battery bank designed to accommodate an occasional extended period of no sun will be sized large enough to necessitate only a shallow level of discharge on a normal, daily basis.

Of course, deep-discharge batteries may be used in both deep- and shallow-discharge applications. When deep-discharge batteries are used in a shallow-discharge application, very long term service can result. The only reason they are not always chosen over shallow-discharge models is the considerable cost difference between the two. With either type of battery, a stratified electrolyte condition may develop unless the electrolyte is periodically mixed by frequent cycling, by the use of a mechanical device, or by a short period of overcharging (equalizing charge).

The location of the photovoltaic system is a very significant factor in battery selection. If the site is remote and unattended, it is very important to have low-maintenance or maintenance-free batteries. High-maintenance batteries should be selected only if proper precautions are taken and the site is accessible so that routine maintenance can be performed easily.

Batteries are usually heavy, and large batteries are very heavy (up to 350 pounds per cell and more). Thus, proper handling equipment must be available at sites where large batteries are used. Because of this logistical problem, very remote systems located in hard-to-access places will often be better served by a large number of small batteries than by a few large batteries.

The size and type of system figure into such considerations. The smaller the system, the more likely that the owner is willing to replace his battery every three to five years. The owner of a large industrial system will usually look for a battery system with a much longer lifetime.

Although most battery manufacturers agree that increasing the number of parallel connections in a battery bank will not reduce reliability, a large number of connections may increase installation costs. By contrast, some manufacturers have stated that a large number of series-connected batteries can reduce reliability. Solar Design Associates has had experience with the use of up to 140 series-connected batteries (14 strings of 10 batteries each) without encountering any reliability problems. However, it is worth emphasizing that the use of proper hardware and professional installation procedures is important for preventing problems.

When systems are installed in remote areas and/or when accurate local climatic data is not available, a conservative value for no-sun days should be selected because of the variability in local climates from year to year. Designing around a conservative number of no-sun days will of course increase the size of the battery storage bank. But this result can be considered desirable, particularly in critical applications such as communications systems.

When a backup power supply such as a diesel generator is available, it is possible to select a shorter period of no-sun days without sacrificing system reliability. The advantages of a backup generator are most apparent in systems that would otherwise have a long (15 days or more) no-sun period. For example, if a mountaintop system requires 20 days of storage, use of a reliable backup generator would allow you to shorten the period to 3 to 5 days. Just keep in mind that most generators must be "exercised" periodically to ensure that they start when called upon.

With batteries, as with most other things, "you get what you pay for." If keeping the initial cost of a system low is a prime consideration, it may be best to use less costly but shorter-life batteries. On the other hand, if life-cycle costs are considered, it may be more economical to use a more costly but longer-life battery. (Table 4-3, found at the end of the discussion of batteries, compares the initial and life-cycle costs of several batteries.)

There are a lot of questions that need to be asked when selecting a battery. How experienced is the manufacturer, how capable and committed to good design? Is this battery a specially designed photovoltaic battery or is it simply a car battery with a new label? How are terminal connections made? What accessories are available? How is the product warranted? Before choosing any particular battery, the PV systems designer should consult and be fully familiar with the manufacturer's design criteria, specifications, and recommended operating and maintenance procedures.

Battery Sizing

When sizing a photovoltaic system to meet a given load, efficiencies (energy output divided by energy input) of all components must be determined. However, when the ampere-hour method is used to size a battery bank, the energy efficiency of the battery is not considered in the sizing calculation. The battery's internal losses are instead considered as another addition to the array's total load. Later, the PV array size and output must be checked to ensure that enough net power is available to satisfy the load after all internal system losses.

Coulombic efficiency—**the ratio between amp-hours entering and amp-hours leaving a battery**

Voltaic efficiency—**the average voltage output of a battery divided by its average voltage input**

Thus, when sizing the battery bank, we consider the amp-hour efficiency (called *coulombic efficiency*) of new batteries to be close to 100 percent, so that amp-hours out equal amp-hours in. Of course, this is not the same as saying that the *energy* efficiency is 100 percent.

What actually happens during the charge/discharge cycle is that a battery is charged by receiving a certain number of amp-hours at an input voltage (for example, 13 to 15 volts for a 12-volt battery) and delivering the same number of amp-hours out at an output voltage that is lower (for example, 12 volts). Hence, a battery's energy efficiency can be calculated as its average voltage output divided by the average voltage input. This is called the battery's *voltaic efficiency* and is what people are most often referring to when they speak of the efficiency of storage batteries incorporated into a photovoltaic system.

The first step in determining the size of the battery required is to calculate the daily load in terms of ampere-hours and the system voltage—as explained in the section on system sizing in chapter 3. The daily load requirements will determine the necessary battery bank capacity. The system voltage will determine the battery bank voltage and the number of cells to be connected in series. In determining daily load, be sure to include all internal system losses (such as inverters).

Next, decide on the number of no-sun days for which to design battery storage. The number will vary according to local weather conditions and the required reliability of the system. (In a few cases, local codes specify the storage requirement.) For example, table 4-2 shows the variation in storage requirement needed in Tucson, Arizona, based on the degree of reliability demanded by the system. This can serve as a model for photovoltaic systems in regions known to have very clear skies almost every day.

For areas with less daily sun (such as Massachusetts), the required amount of storage is two to three times as high. Photovoltaic system engineers have developed computer sizing programs that use a statistical reliability analysis to determine the optimum number of days of storage for a required level of reliability in a given system.

Table 4-2. System Reliability and No-Sun Day Design Period in Tucson, Arizona

Application	Required Reliability	No-Sun Days
Lighting Water pump with storage Appliances	Medium 90–95%	3
Cathodic protection at remote site	High 95–98%	5
Communications at "inaccessible" sites	Very high Better than 98%*	8

*It is unrealistic to consider a system that is 100% reliable.

To calculate the *total usable battery capacity* (TUC) required by a system, multiply the daily load by the number of no-sun days of storage required. (This can be expressed in the formula TUC = DL × N.) It is important to remember that a system's total usable battery capacity is not the same as the nominal rating of the battery capacity as specified by the manufacturer, which is simply designated C.

Once the system's required TUC has been determined, the necessary individual battery capacity can be calculated and the proper number of batteries specified. The calculation procedure differs according to the design and performance of the particular battery. The parameters most often included in the manufacturer's product specifications are daily depth of discharge, maximum allowable depth of discharge, rate of charge and discharge, operating temperature, and the desired capacity remaining at the end of the battery's useful design life.

The following procedure is recommended for calculating the amount of the rated capacity of the individual battery under consideration that will be available for use and then determining the TUC and the number of batteries required to satisfy the storage requirements of a particular PVsystem. Again, we will use the Delco 2000 as an example.

Step 1: Determine the system's voltage and daily load and the number of no-sun days of storage required.

Step 2: Study the manufacturer's recommendations on capacity adjustments for battery operating temperature as well as charge and discharge rates and de-rate the batteries to fit the proposed system. (The available rated capacity of a Delco 2000 as a function of operating temperature and discharge rate is illustrated in figure 4-4.)

Step 3: To determine the TUC for the individual battery being considered, multiply the available capacity determined in the previous step by the manufacturer's recommended MDOD. This will ensure that the battery is not discharged more than its maximum even at the end of the no-sun day design period. (The recommended MDOD for the Delco 2000 will be considered as 50 percent for this exercise.)

Step 4: To determine the number of batteries required in parallel (B_p) to satisfy the system's storage requirements, divide the system's TUC storage requirements (previously described as DL × N) by the available TUC of the individual battery, the product of the de-rated capacity and MDOD obtained in Steps 2 and 3. This procedure can be synthesized into the following formula:

$$B_p = \frac{\text{System TUC}}{\text{Battery TUC}} = \frac{DL \times N}{MDOD \times C}$$

If the number of batteries in parallel calculated is not a whole number, always round up.

Step 5: To find the number of batteries required in series (B_s) to satisfy the system's storage requirements, divide the system voltage (V) by the nominal battery voltage (V_{bn}). Since the nominal voltage of the Delco 2000 is 12 volts, the formula in this case will be $B_s = V \div 12$.

With batteries in series, in contrast to batteries in parallel, you cannot round up or down, since you will very likely deviate too far from the "voltage window" acceptable to the load. Thus, if the number of batteries in series calculated is not a whole number, the system's voltage must be changed, since you cannot purchase a fraction of a battery. If you cannot change the voltage of the system, you must either consider using a different, lower-voltage battery or specify a DC-to-DC converter between the battery and the load.

Step 6: The total number of batteries (B_t) required to satisfy the system's storage requirements is the product of the number of batteries in parallel and the number of batteries in series ($B_t = B_p \times B_s$).

Sample Exercise in Battery Sizing

In order to illustrate the battery sizing process just described, let's see how it would be done for a specific system. Our example will be a simple, stand-alone photovoltaic system located in the southwest part of the United States.

Step 1: First we need to establish our basic system design parameters. They are as follows.

- System load: 42.4 Ah/day
- System voltage: 48 VDC
- System location: Fish Creek, Arizona
- Number of no-sun days: 4 (based on the fact that operation is important, but not critical)

Step 2: Next we need to determine the rated battery capacity, which we will do with the Delco 2000 as our example. It is assumed that the battery will be used for about 8 hours per day and will be kept near 21°C (70°F) during the winter months. If 42.4 amp-hours are being drawn over 8 hours, the average discharge rate is 42.4 divided by 8, or 5.3 amps per hour. Consulting the manufacturer's data for the Delco 2000 presented in figure 4-4, we find that the capacity of the battery is approximately 96 amp-hours for this discharge rate and this operating temperature.

Step 3: Now let's determine the individual battery TUC by multiplying the capacity of 96 amp-hours by the manufacturer's recommended maximum depth of discharge, which we will consider to be 50 percent for the Delco 2000: 96 Ah × 0.5 = 48 Ah.

Step 4: We can now calculate the required B_p by dividing the TUC requirements of the entire system (DL × N) by the individual battery TUC.

$$B_p = \frac{42.4 \times 4}{48} = \frac{169.6}{48} = 3.53$$

Since we cannot use a fraction of a battery, we round 3.53 off to 4 batteries.

Step 5: The DDOD should always be checked to see if it is within the manufacturer's recommended limits. We have in this case a total amp-hour storage capacity of 4 (batteries) times 96 amp-hours, or 384 amp-hours, and we are withdrawing 42.4 amp-hours each day. Therefore, the DDOD is 42.4 divided by 384, or 11 percent. This is less than the DDOD recommended by Delco, which is 15 percent. By keeping within the manufacturer's recommendations, this battery bank can be expected to have a life of about 1,700 cycles, or nearly five years.

Step 6: We now calculate the B_s using the previously explained formula, $B_s = V \div V_{bn}$. Since the system voltage was established in Step 1 as 48 VDC and the nominal voltage of the Delco 2000 is 12 volts, the calculation here is quite simple: $B_s = 48 \div 12 = 4$.

Step 7: Finally, we determine B_t by multiplying the B_p (4) by the B_s (4). Thus, 16 batteries are required to complete the system.

Alternate Sizing Calculation

Before leaving our simple system example, let's consider for a moment how the numbers would change if we altered one of our basic design parameters. Let's suppose we decided at the outset that the level of reliability required by this particular system was low enough that we would need to provide only two days of storage, rather than four. This would alter our calculation of the number of batteries needed in parallel.

$$B_p = \frac{42.4 \times 2}{48} = \frac{84.8}{48} = 1.77$$

Rounded off, the result would indicate the need for only two batteries in parallel. This would result in a total amp-hour capacity of 2 (batteries) times 96 amp-hours, or 192 amp-hours. Since we are taking out 42.4 amp-hours each day, the DDOD would increase to 22 percent ($42.4 \div 192 = 0.22$).

In this case, we exceed the manufacturer's recommended allowable 15 percent DDOD. Consequently, making this change would shorten battery life to about 700 cycles (less than two years). Since such a short battery life is probably not acceptable, either the number of batteries should be increased or a battery that allows a deeper daily discharge should be specified for this system.

Interpreting the Comparison Table

A comparison among some of the various batteries currently used in photovoltaic systems is provided in table 4-3. Here are some observations to help you interpret and make use of that information.

The figures found in the column labeled "List Price" refer to the approximate cost when small quantities of batteries are purchased on a retail basis. The actual cost may be less when large purchases are made.

Note that some battery models appear twice in the table. These models are provided with separate ratings by their manufacturers, depending on whether they are to be used as deep- or shallow-cycle batteries. By comparing the two sets of data, you can see how daily depth of discharge affects battery life and the ultimate system cost.

Performance characteristics cited in the table were all obtained from manufacturers' literature and should not be considered measured performance values.

In the table, total kilowatt-hours (TkWh) delivered are equal to the DDOD times the V_{bn} times the number of cycles times the amp-hour rating divided by 1,000. The cost per kilowatt-hour of electricity delivered from storage is then calculated by dividing the list price by the total number of kilowatt-hours. You will note that the costs of transportation and installation, both important considerations when selecting a battery, are not factored into the cost per kilowatt-hour as presented here, since they are specific to the individual system.

It is very important to consider the system's specific parameters when determining life-cycle costing and selecting a battery. They include the cost of freight, handling, and labor, both for the initial installation and for later replacements. As an example, we see from table 4-3 that the Delco 2000 is an inexpensive battery, but it must be replaced every two years if discharged at a rate of 20 percent per day. By contrast, the Exide Tubular Modular battery will last ten years at a 20 percent DDOD. As a result, the life-cycle cost of the Exide may be considerably less than that of the Delco. This especially might be true if the batteries were installed in a very remote location where transportation and installation costs for frequent replacement might be considerable.

Battery Chargers

There is an additional BOS component that serves a vital purpose in a stand-alone system between the auxiliary generator and the battery storage bank. This is the *battery charger,* a device that converts the AC output of the generator into DC electricity acceptable for storage in the batteries. A battery charger is included in a photovoltaic system with a back-up power source to recharge the battery bank in the event that the batteries reach their recommended MDOD and array input is not sufficient to recharge them. The battery charger also supplies the necessary equalizing charge that prevents stratification of the batteries' electrolyte solution.

A serviceable battery charger intended for use with an SA photovoltaic system should have several major features. First, it should be specifically designed to work with an on-site, fuel-fired generator and be capable of deep-cycle battery charging. Second, it should have an output

continued on page 116

Table 4-3. Comparison of Various Storage Batteries

Manufacturer and Model	Model Number	List Price ($)	Shallow/ Deep Cycle (S/D)	Nominal Capacity (Ah)*
GNB Absolyte	638	63	S	42
	1260	114	S	59
	6-35A09	654	S	202
	3-75A25	1,168	S	1,300
Exide Tubular Modular	6E95-5	600	S	192
	6E120-9	930	S	538
	3E120-21	945	S	1,346
Delco-Remy Photovoltaic	2000	78	S	105
Globe Solar Reserve Gel Cell	3SRC-125G	181	S	125
	SRC-250G	100	S	250
	SRC-375G	150	S	375
Globe	GC12-800-38	100	S D	80 80
GNB Absolyte	638	63	D	40
	1260	114	D	56
	6-35A09	654	D	185
	3-75A25	1,168	D	1,190

Nominal Voltage (V)	Daily Depth of Discharge (%)	Life (Cycles)	Life (Years)	Number of Sets over 20-Year PV Lifetime	Total Power Delivered (kWh)	Cost ($/kWh)
6	50	1,000	2.7	8	126	0.50
12	50	1,000	2.7	8	359	0.32
12	50	3,000	8.0	2	3,636	0.18
6	50	3,000	8.0	2	11,700	0.10
12	15	4,100	10.0**	2	1,417	0.42
	20	3,900	10.0**	2	1,797	0.33
12	15	4,100	10.0**	2	3,970	0.23
	20	3,900	10.0**	2	5,036	0.19
6	15	4,100	10.0**	2	4,967	0.19
	20	3,900	10.0**	2	6,299	0.15
12	10	1,800	4.9	4	227	0.34
	15	1,250	3.4	6	236	0.33
	20	850	2.3	8	214	0.36
6	10	2,000	5.5	4	150	1.21
2	10	2,000	5.5	4	100	1.00
2	10	2,000	5.5	4	150	1.00
12	20	1,500	4.0	5	288	0.35
12	80	250	less than 1.0	many	240	0.42
6	80	500	1.4	14	96	0.65
12	80	500	1.4	14	269	0.42
12	80	1,500	4.0	5	2,664	0.25
6	80	1,500	4.0	5	8,568	0.14

continued

Table 4-3. Comparison of Various Storage Batteries—*continued*

Manufacturer and Model	Model Number	List Price ($)	Shallow/ Deep Cycle (S/D)	Nominal Capacity (Ah)*
Surrette	CH-375	231	D	375
	NS-29	522	D	490
	NS-33	813	D	564
Exide Tubular Modular	6E95-5	600	D	180
	6E120-9	930	D	360
	3E120-21	945	D	1,250

NOTE: Data compiled in March 1985.
*Amp-hour rating taken at 5-day rate for shallow cycling and 24-hour rate for deep cycling (when data was available from manufacturer).
**Limited by manufacturer.

curve that is "tapered" to favor the initial phase of the charging cycle while trickling off gently as the batteries approach full capacity. Third, it should feature a high/low output switch, so that output may be increased if the battery bank is later enlarged. Finally, it must be capable of functioning in constant interface with the generator, with its switch always in the "on" position for immediate activation and charging.

The battery charger linked to an auxiliary generator does its work in a relatively simple fashion. Whenever it is necessary for the generator to kick in, the battery charger also comes on. (This is always true, except possibly on those occasions when the generator is activated in its "exercise" mode, as explained in chapter 5.) As a state of full charge is approached, the charger automatically tapers the input current to apply a finishing charge at a slower rate. This prevents overcharging in a manner similar to that in which a quality voltage regulator controls the charging current from the photovoltaic array.

The equalizing charge is done in a different way. Its purpose is to destratify the electrolyte, and this is done when the batteries are already at a state of full charge. In this case the batteries are carefully and deliberately overcharged, with a manual control employed in all but the most sophisticated systems to terminate the process at just the right time.

Those are the general guidelines for selecting a battery charger. When sizing the unit, the following rules apply:

- To prevent generator overloading, make sure the output amperage of the generator exceeds the amperage draw of the charger

Nominal Voltage (V)	Daily Depth of Discharge (%)	Life (Cycles)	Life (Years)	Number of Sets over 20-Year PV Lifetime	Total Power Delivered (kWh)	Cost ($/kWh)
6	80	1,400	3.8	5	2,520	0.09
6	80	1,400	3.8	5	3,293	0.16
8	80	1,400	3.8	5	5,053	0.16
12	80	1,800	4.9	4	3,110	0.19
12	80	1,800	4.9	4	6,221	0.15
6	80	1,800	4.9	4	10,800	0.09

and the house AC loads that the generator will be expected to power by at least 10 percent.

- Make sure that voltages are compatible. If your generator puts out both 120 and 240 VAC, choose a charger with 240 VAC input to reduce the amperage draw on the generator and prevent damage from overheating. On the other hand, make sure that the output voltage of the charger is compatible with the battery bank voltage, which will be identical to whatever DC line voltage is used for residential loads.
- Select a charger with a maximum hourly charge rate in amps that is between 15 and 25 percent of the battery bank amp-hour capacity. A charger of higher than 30 percent capacity would not be fully utilized because higher charge rates will not restore full charge, and damage to the batteries could result from frequent rapid recharging. This is because a battery's ability to receive charge is inversely proportional to the rate of charge.
- Select a charger with a low start-up surge (approximately 25 percent of the maximum rated input current of the charger) so that the generator will not be unduly taxed initially during cold starts. Or install a timer that delays connection of the battery charger until the generator is up to temperature.
- Select a charger model compatible with the particular type of battery you are using in your system. For example, if you use lead-acid batteries (by far the most common type in photovoltaic

installations), make sure the charger is compatible. If you use nickel-cadmium or some other type, the same rule applies.

Adherence to these sizing standards, and to the specific recommendations of the battery charger manufacturer, will ensure longer battery life and more efficient operation of the auxiliary power source—both primary considerations in a smooth-running stand-alone photovoltaic installation.

This chapter has focused attention on the balance-of-system components necessary for delivering photovoltaic-generated DC electricity to loads in a stand-alone system. The following chapter describes the sophisticated hardware that turns DC electricity into AC electricity, while also discussing options for backup generating and system monitoring and control devices. As you will see, several of the components and design guidelines described for stand-alone systems have their place in utility-interactive AC systems as well.

CHAPTER FIVE

INVERTERS, BACKUP GENERATORS, AND SYSTEM CONTROLS

A photovoltaic (PV) array, regardless of its size or sophistication, can generate only direct current (DC) electricity. Fortunately, there are many applications for which direct current is perfectly suitable. Even more fortunately, DC electricity can be converted to alternating current (AC) with relative ease and efficiency through the use of a piece of equipment called an *inverter*. It is the inverter that makes PV technology compatible with the type of equipment and appliances encountered in the average home.

***Inverter*—an electronic device used to convert direct current electricity into alternating current**

There are two possible input sources of photovoltaic electricity in a house: the array itself and (in a stand-alone system) the battery storage bank. If the load demand is for alternating current, an inverter must occupy a position in the system between the array and the load or between the batteries and load.

Inverters are nothing new. They have been around as long as there has been a need for converting DC into AC electricity. The early *rotary* type of inverter had internal moving parts. The DC electrical source powered a DC motor connected to an AC alternator, which produced AC electricity for the load. Rotary inverters are still manufactured, largely for use in marine and aircraft electrical systems where a clean AC signal is desired and efficiency is not critical.

Virtually all the inverters used with alternative power systems are transistorized, solid-state devices. Solid-state inverters are preferred for their higher efficiency, ease of maintenance, and infrequency of repair. Broadly speaking, these inverters may be divided into two categories: stand-alone (SA) and utility-interactive (UI), or line-tied. The stand-alone inverter is capable of functioning independently as part of a home electrical generating system unconnected to the public utility grid. It is activated solely by the incoming DC power from the photovoltaic array (or batteries, or wind turbine, or hydroelectric generator). The utility-interactive inverter, on the other hand, is compatible only with a PV power system that is tied to the utility grid.

PHOTO 5-1

This high-efficiency DC-to-AC inverter, manufactured by Dynamote Corporation for stand-alone applications, delivers a pure sine wave output.

Hertz—a unit of frequency equal to 1 cycle per second

In both SA and UI photovoltaic systems, the inverter must do more than simply change DC into AC electricity. It also must "condition" the array output so that it can efficiently operate the various AC components of the load—appliances, electronic devices, and so on. Utility-interactive inverters have the added task of integrating smoothly with the current, voltage, and frequency characteristics of the utility-generated power present on the distribution line. In the United States, the polarity of the AC signal changes 120 times per second and its frequency is called 60 cycle or 60 hertz (Hz). In Europe and many other foreign countries, the polarity of the AC signal changes 100 times per second and it's called 50 cycle or 50 hertz.

An inverter is expected to provide a stable voltage output, a steady AC frequency, and a wave form that departs as little as possible from the basic sinusoidal wave shape of the AC sine wave. Inverter conversion efficiency is also a very important consideration. Internal losses in a DC-to-AC inverter can consume between 5 and 40 percent (or more at low power levels) of the system's DC output power.

In SA applications, sizing of the inverter is very critical. The unit must be large enough to handle motor-starting surge inrush currents and

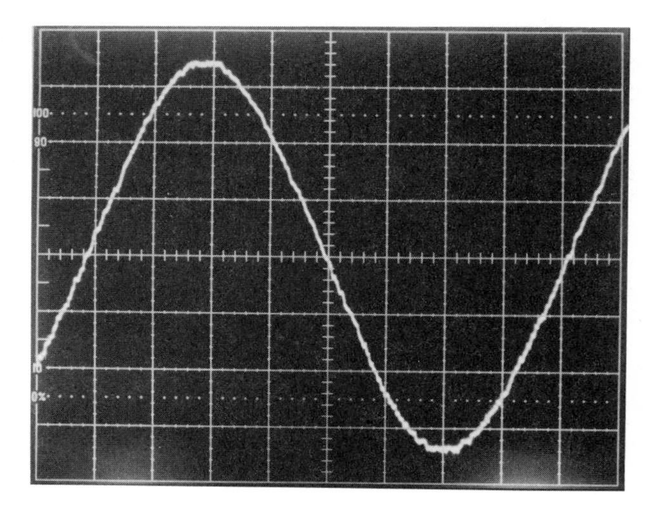

PHOTO 5-2

An oscilloscope photo of the sine wave produced by the Dynamote UXB stand-alone inverter.

the resultant short-duration peak loads if two pieces of equipment happen to start up at the same time. However, care must be taken to avoid oversizing the unit because it will not deliver its peak efficiency when operated at only a portion of its rated capacity.

The Utility-Interactive Inverter

The UI inverter differs from the SA unit in that it can function only when tied to the utility grid. In effect, it provides the interface between the photovoltaic array and the utility, and makes possible the sell-back of power provided for under the Public Utility Regulatory Policies Act (PURPA) of 1978.

The UI inverter not only conditions the power output of the photovoltaic array, it also serves as the system's control and the means through which the site-generated electricity enters the utility lines. The UI inverter uses the prevailing line-voltage frequency on the utility line as a control parameter to ensure that the PV system's output is fully synchronized with the utility power.

***Utility-interactive inverter*—an inverter compatible only with a photovoltaic system tied to a utility grid**

It is worth noting that in many UI photovoltaic systems much if not most of the PV array output is sold to the utility rather than used directly on-site, since the timing of peak array output and peak load requirements seldom coincide. Because of this situation, the utility rate structure should be examined and factored in to the system planning for a PV installation of this type.

Utility-interactive inverters are not bound by the surge-capability requirements of SA units because the utility will satisfy any load surge requirements, but sizing is very important nonetheless. The inverter must be carefully chosen to match both the photovoltaic array output and the utility interface parameters. One or two isolated line-tied PV systems are unlikely to cause much difficulty. However, a sizable aggregate of badly designed or poorly functioning inverters could cause variations in wave form and frequency on the distribution line beyond the utility's acceptable limits.

At first glance, a UI photovoltaic installation looks much the same as a stand-alone from the outside. However, there are fundamental differences in the internal relationship of the components, particularly with regard to the function of the inverter. A UI system is not just an SA system without batteries.

In a line-tied system, DC electricity generated within the modules of the PV array flows through lightning protection and source combining equipment directly to the inverter, which (along with its internal controls and filters) is referred to as the *power conditioner.* At the power conditioner, the raw DC output of the array is converted to AC electricity of the quality required for load consumption and grid interface. Then, depending upon the level of output and the fluctuating demand of the house loads, the AC power flows either to the loads alone or to both the loads and the utility grid.

***Power conditioner*—the integrated assembly of inverter, inverter controls, and filters that make DC array output in a utility-interactive photovoltaic system usable for AC load application and grid interface**

PHOTO 5-3
State-of-the-art, 2- and 4-kilowatt utility-interactive inverters (center) and string combiner boxes manufactured by American Power Conversion Corporation.

With conventional equipment, the UI photovoltaic system cannot operate without the presence of utility power. If you desire to draw power from your utility-interactive PV system during a utility outage, you must either install a second inverter capable of stand-alone operation or one of the new, semiexperimental, bimodal, self-commutated inverters capable of either line-tied or SA operation.

At present all utilities require that photovoltaic systems immediately disconnect from the grid whenever the grid power fails. This is to eliminate any possibility of PV-generated power being fed out onto a downed line and endangering service personnel. The new inverters are not likely to cause a change in policy by the utility companies unless and until it can be satisfactorily proved that they will not send current out onto the lines when the grid is down.

Since there can be no photovoltaic contribution at night, most higher-quality UI systems incorporate an automatic "night-mode" sensing circuit, which deactivates the inverter during periods when PV-generated power is very low. This feature is desirable because a line-tied inverter kept in operation at night by incoming grid current, but doing no work, can represent a large parasitic load in and of itself.

Much is asked of a UI inverter, both by the system itself and by the utility to whose grid it will feed power. Some of these requirements will vary with the size of the installation, while others are implicit in the nature of the demands placed upon grid-interfaced dispersed power producers and will apply in all cases. These latter requisites are now assumed to be fundamental by all manufacturers of quality UI inverters and should not have to be individually specified by the photovoltaic system designer. Nevertheless, it's a good idea to have at least a general familiarity with this piece of equipment—what it's supposed to do and why.

Selecting a Utility-Interactive Inverter

There are many different factors that affect the cost and performance of a UI inverter. *Size* and *capacity* are two important variables. The

UI inverter has to be sized large enough to handle the photovoltaic array output power, including the higher-than-normal *source surges* that are generated during certain weather conditions and produce up to 1.4 times the normal solar insolation intensity (1.4 kWh/m^2) on the array by magnification from cloud and snow reflections. All quality UI inverters are designed to handle these occasional periods of increased array output. If this is not clear from the manufacturer's literature, you should inquire about the capabilities of a particular unit before you purchase it. Load surges such as those from motor-starting inrush currents, which can be troublesome for SA inverters, are not an important issue in UI systems, since the utility grid itself will always cover the home's peak requirements.

Most UI inverters achieve their *maximum operating efficiency* at or near full-power rating. Since the solar insolation intensity and the angle of incidence on the array as well as temperature and other operating parameters of the PV system are constantly changing, the inverter operates at its full-power rating only a small portion of the time. It is, therefore, very important to choose an inverter with a *high average efficiency* or, ideally, a (nearly) flat efficiency curve.

When comparing efficiency ratings in manufacturer's literature, be aware that the inverter manufacturer may have a somewhat overly optimistic view of its product. Responsible manufacturers publish the output efficiency of their units as a function of the percentage of rated power from one-eighth power to full load. Many provide this information in graphic form as well. If this information isn't available or easily understood for the unit you are considering, inquire about it. If possible, obtain actual test data developed by an independent third party where the test conditions and resulting assessments are likely to be more objective.

Figure 5-1 presents a family of efficiency curves for a high-quality, 4-kilowatt UI inverter manufactured by American Power Conversion Corporation. The three curves—LH, NN, and NL—show the effect of variations in source and utility-line voltage on inverter efficiency. LH designates a low DC input voltage of 207.1 volts and a high utility voltage of

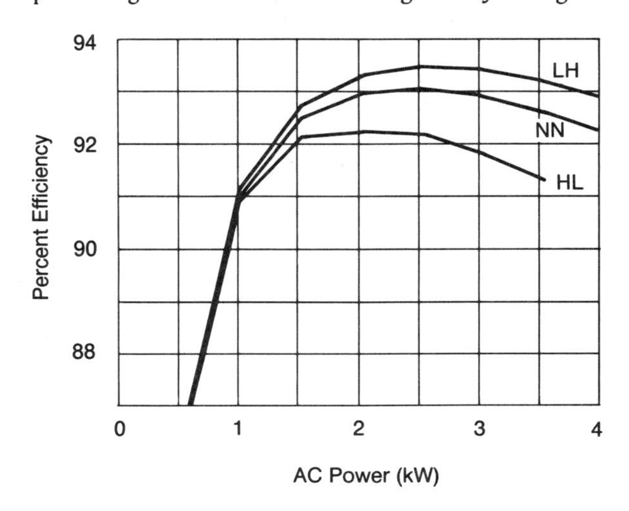

FIGURE 5-1

Efficiency curves for the 4-kilowatt UI-4000 utility-interactive inverter manufactured by American Power Conversion Company.

254.6 volts. The NN curve designates a normal condition of 230.2 volts of direct current (VDC) input and 240 volts of alternating current (VAC) line. The HL designates a high DC input of 270 volts and a low AC line of 209 volts.

Maximum-power-point tracking—the ability of an inverter at any given time to locate and operate at the voltage that maximizes the PV array output

The importance of inverter efficiency as a significant criterion in photovoltaic system design cannot be overstated. An increase of a percent or two in average efficiency could mean many thousands of kilowatt-hours of increased output over the life of your PV system. To compensate for the constant variations in a system's output characteristics, many of today's more sophisticated inverters are able to perform *maximum-power-point tracking*. This is the ability of an inverter to find and operate at the array voltage that maximizes the output power of the PV array at any given instance, under any set of conditions.

Most high-quality UI inverters have an electronic control circuit that enables them to determine the voltage that offers the highest power level on a continuing basis, thus enabling the system to make the most of prevailing circumstances. This is loosely analogous to the automatic exposure setting on a camera, where either aperture or speed is automatically adjusted according to the light conditions to give the best exposure.

Grid-Interface Issues

Inverter efficiency and maximum-power-point tracking are of direct concern to the owner of a UI photovoltaic system. However, there are a variety of other factors of principal concern to utility companies that the owner or designer of a line-tied PV system must be aware of. These factors have a potential effect upon the quality of the power in the utility grid and thus upon policies governing the acceptability of UI arrangements.

Until now, most of these concerns have remained largely academic, since there are hardly enough photovoltaic, wind, or other alternative generating facilities interacting with the utility grid to cause any real problems. However, as dispersed power generation becomes more widespread and the sources of power signals more numerous and diffuse, each of these influences on signal quality will grow in importance. Code authorities and the utility industry have already realized the importance of establishing uniform standards early in an attempt to qualify and avoid problems before they happen. Much has been accomplished to date.

The 1984 version of the *National Electrical Code* (NEC) has a new section, Article 690, devoted exclusively to photovoltaic systems, and Underwriters Laboratories (UL) has developed a set of standards for UI inverters. Also, the Institute of Electrical and Electronics Engineers (IEEE) is working with the Electric Power Research Institute (EPRI) and other utility representatives to develop a universal set of standards on power quality and safety for the grid-interconnection of PV, wind, and other dispersed energy systems.

Sound and radio signals travel in waves. So does AC electricity, and the *wave shape* of the AC signal present on the utility power line is important. The ideal wave shape for AC current is the *sine wave* that is the

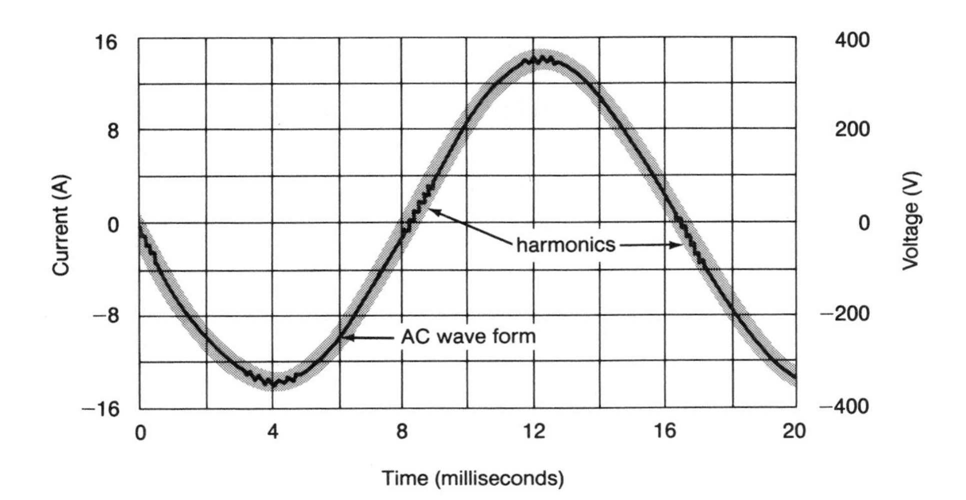

FIGURE 5-2

A diagram of the current and voltage output wave forms of a utility-interactive inverter, showing harmonic distortion present in the current wave form.

standard on the utility grid. In order for a UI inverter to have an interactive relationship of any duration with the utility grid, the inverter's output must, by definition, match or closely approximate the utility-supplied wave shape. The degree to which the AC output wave form of inverters resembles that ideal varies from model to model.

In all solid-state inverters, the conversion of direct current to alternating current is accomplished through the use of an electronic switching circuit. Generally speaking, the higher (faster) the switching frequency, the higher the quality of the resulting AC wave form. Since the utility-generated signal alternates at 60 cycles per second, 60 hertz is the slowest switching frequency that could work effectively. Early line-tied inverter designs were line-commutated and relied on the utility-supplied 60-cycle signal to drive their internal switching circuitry. They generally employed silicon-controlled rectifiers (SCRs) as power semiconductors and were reliable and reasonable in cost.

However, the signal quality of these early units contained a very high (up to 30 percent) harmonic distortion and delivered a rather unimpressive power factor (from 0.3 to 0.7 lagging depending on DC voltage and power level), and thus left a lot to be desired. Recent developments in advanced thyristors, bipolar silicon transistors, and power field-effect transistors have made it possible for today's state-of-the-art inverters to employ much higher—up to 100 kilohertz (kHz)—internal switching frequencies and produce a nearly perfect sine wave with very little harmonic content that is quite acceptable to the utilities and harmless to household induction motors.

Kilohertz (kHz)—AC frequency unit of measurement equal to 1,000 cycles per second

Harmonics are closely related to wave shape. The harmonic content of either the voltage or current wave form is a measurement of the secondary signals and other interference present, which when taken in total with the primary sine wave signal either adds to or subtracts from this primary signal, resulting in a distorted wave form, as shown in figure 5-2.

Harmonics—the amount of departure from the sine wave shape of an AC wave form

Total harmonic distortion (THD)—the total amount of secondary signals and other interference that distorts the wave form of a primary signal

Power factor—a qualitative measure of the degree to which the voltage and current wave forms in an AC signal are in phase

If the *total harmonic distortion* (THD) goes much beyond 15 or 20 percent, an appreciable increase in the operating temperatures of induction motors becomes apparent. As an example, a square wave, which is supplied by low-quality SA inverters, has a THD of 48 percent. The better line-tied inverters presently available are very low in current harmonic distortion, generally below 5 percent. Utility-interactive inverter technology has developed rapidly in recent times and the very best contemporary designs are actually capable of producing a "cleaner" AC signal than that which normally prevails on the utility grid itself.

Power factor is a qualitative measure of the in-phase relationship between the voltage wave form and the current wave form, which together make up the AC signal in the utility grid. The ideal power factor, called the *unity power factor,* is achieved when both wave forms occur in precise harmony and is represented by the value 1.0. Departures from unity power factor are indicated as fractions of that optimal number. A low power factor results when these wave forms are out of phase and is indicative of the amount of reactive or "parasitic" power in the signal being sent through the lines.

Inductive electric motors—such as those found in refrigerators, freezers, other large appliances, and industrial equipment—can bring down the power factor across the line. Utility-interactive inverters can also have a negative effect on power factor. Some of the early designs put out a signal with a power factor rating of as low as 0.3. The top-of-the-line products currently being sold, however, often have a power factor rating approaching 0.95 and are quite capable of meeting the utility's requirements. These requirements are now beginning to be applied as a general policy even though it will be a matter of many years before enough UI inverters are present on the nation's grids to significantly affect overall signal quality.

Voltage stability is another important concern. Voltage harmonic distortion in a UI system is the result of harmonic current from the inverter acting on the utility voltage. It is very unlikely that this problem will occur in any but the largest photovoltaic systems, especially if the inverter's current harmonic distortion is kept below 5 percent. Voltage regulation in a utility-interactive PV system will nearly always be determined by the utility's generation source. This is because the utility's source is so many times larger than the PV array's output that the UI inverter has no choice but to work at the utility-supplied voltage (or not work at all). All UI inverters must be capable of operating within the normal operating range or "window" of utility-supplied voltage, which is specified by the American National Standards Institute (ANSI) to be between 216 and 254 volts for a typical 240-VAC service.

The *frequency stability* of the AC cycle is important because all synchronous motors respond and lock into the frequency of the power supply on which they are drawing. Any departure from the proper cycle can cause erratic operation of equipment such as clocks and phonograph turntables. In a UI photovoltaic system, deviations from the appropriate 60 hertz frequency are unlikely to be a problem because the inverter itself must be synchronized with the grid-supplied signal in order to operate. Of course, this is not the case in stand-alone PV systems where the inverter must generate and control the frequency of its AC output.

Electromagnetic interference (EMI) is the line-transmitted crackle in radio or TV reception that sometimes occurs when another appliance is switched on nearby. Higher-frequency EMI called *radio-frequency interference* (RFI) is actually transmitted from the offending equipment through the air much like radio and TV signals.

Inverters (both line-tied and stand-alone) can themselves be a source of both high- and low-frequency EMI, affecting not only the homes where they are installed but also other sites close by. Television and radio reception (especially AM), telephone conversations, and the operation of computers can all be affected by EMI and RFI. This problem, which seems to be more of a vexation with SA inverters than with UI units, has been recognized for some time. Manufacturers of quality inverters have equipped their devices with internal filter circuits capable of screening out this unwanted interference. Such modifications detract slightly from efficiency and increase costs, but they help ensure that the inverters will not be a source of either utility consternation or neighborhood annoyance.

Electromagnetic interference (EMI)—distortion in the operation of one electronic device caused by interference from a second device located nearby

Inverters also make audible noise when they do their work, though the amount can vary considerably from model to model, depending on the basic circuit design as well as the packaging of the unit. Some units are markedly quiet whereas others are downright disturbing. Noise is also proportional to the load: an inverter will make the most noise under full load. When selecting your inverter, be certain you know how much audible noise it will produce under load, then mount it away from noise-sensitive areas.

There is also the matter of *safety.* When there is a blackout on any portion of the power grid, the utility must be able to rely on the assumption that the lines are completely dead before sending repair personnel out to work on them. If there are no dispersed small power producers along the grid, the utility has no problem in making this assumption. But if there is just one line-tied independent electricity producer continuing to feed power into the downed utility lines, the consequences could be very serious for a lineman working on repairs—particularly if the photovoltaic system's output were to pass through a transformer and be stepped up to 4,000 volts.

If all UI photovoltaic systems were to fall completely silent upon grid failure, the problem would solve itself. But there are situations in which this might not happen. Some of the less sophisticated inverters of early design have, under certain conditions, continued to produce both voltage and current capable of entering the power lines, even when the utility signal is absent. This phenomenon is called *run-on.*

It is also possible, in the event of a utility outage, for two or more line-tied inverters that are feeding power into the same utility distribution feeder to mistake one another's output for the utility's signal and continue to operate. For a more detailed discussion of this phenomenon and some of the solutions being developed, see the U.S. Department of Energy

(DOE) report, *Loss of Line Shutdown for a Line-Commutated Utility Interactive Inverter,* by E. E. Landsman (Massachusetts Institute of Technology, 1981).

A high priority of both inverter manufacturers and utility companies involved with photovoltaic systems is to assure the near-instantaneous shutdown of line-tied inverters in the event of a utility power outage. This is true not merely because of the obvious safety considerations already mentioned. A more immediate concern is that PV systems not interfere with the utility fault-clearing process that occurs immediately after a line fault is detected. For these reasons, all UI inverters must be designed to shut down within milliseconds of a utility power failure and remain off the line until several seconds after normal utility voltage and frequency have been restored to normal functioning.

Another safety-related feature of UI inverters required by the NEC is the grounding of one side of the photovoltaic array. For this and other reasons, the UI inverter must be equipped with an isolation transformer, whose function is to provide a point of electrical isolation between the DC and AC sides of the system. The main purpose of the isolation transformer is to provide a positive means of ensuring that, in the event of an internal fault or failure, DC electricity can not possibly be fed onto the utility line. The majority of UI inverters currently on the market have isolation transformers included in their design.

Photovoltaic system owners and designers should be aware of the utilities' requirements in these crucial areas and must be certain they are met in their choice of inverters. A comparison of some better-known UI inverters is provided in table 5-1.

The Stand-Alone Inverter

While an SA inverter is free of the technical constraints of having to interact and maintain a good relationship with the utility grid, in some respects it has a more difficult job to do. As the name implies, the stand-

Table 5-1.
Comparison of Utility-Interactive Inverters

Inverter	Size (kW)	Overall Efficiency (%)	Current Wave Form	Power Factor	Max-power Tracking
Abacus	4	90	Excellent	1.0	Yes
DECC	4	90–95	Excellent	1.0	Yes
American Power	4	94	Excellent	1.0	Yes

alone unit is completely on its own. Where the line-tied unit has its frequency and operating output voltage established for it by the utility's generation equipment, the SA unit must establish and maintain the frequency and voltage of its own AC output. In addition, the SA unit must be capable of providing (and surviving) the large *load surges* of inrush current required to start many of the common motor loads such as water pumps, power tools, and refrigerator compressors.

Stand-alone inverter—an inverter activated solely by direct current entering it from a photovoltaic array or other DC power source

The SA inverter must also be able to maintain its output voltage wave form while supplying the *out-of-phase currents* that are caused to flow by *inductive loads* such as the power tools just mentioned. This is necessary to avoid disturbing the operation of sensitive electronic equipment such as a personal computer, which might also be running at the same time.

Fundamentally, before a particular inverter is selected, there are two basic system parameters that must be established. The first parameter is the system's DC operating voltage, which will determine the inverter's input voltage. The second is the size of the load both in terms of total watt-hours or kilowatt-hours per day and in terms of the size and type of the larger, individual peak loads that the inverter will be expected to operate.

Let's consider the matter of *input voltage*. There are SA inverters that will take almost any DC voltage and boost it to standard line current. However, inverter efficiency (and capacity) is inversely proportional to the amount of "stepping up" required from the DC input voltage to deliver AC line current. For example, if all other factors are equal, a 120-VDC inverter will deliver 120 VAC at a higher conversion efficiency than a 12- or 24-volt model. In order to obtain the maximum conversion efficiency from the inverter, its DC input voltage should be specified as high as possible.

Specifying a high DC operating voltage will serve to reduce line losses on the input system side of the inverter and eliminate the need for the more expensive, higher-current-handling switch gear. For example, a 5-kilowatt inverter obtaining its DC input from a 120-volt battery bank will draw less than 45 amps at full load. If this same inverter is serviced from a

Transformer Isolation	EMI & RFI	Reliability	Retail Cost ($ approx.)
Yes	Fair	Fair	9,000
Yes	Excellent	Good	8,000
Yes	Excellent	Excellent	7,000

48-volt battery bank, the required current draw is over 100 amps. If 24 volts are used, the figure is over 200 amps.

Nonetheless, taken on its own, an attitude of "the higher the better" is too simple an answer when considering the choice of operating voltage for an SA system. Remember the battery sizing procedure outlined in chapter 4, which stated that batteries must be connected in series to achieve higher voltage and in parallel to achieve higher current? A choice of a very high DC operating voltage for a relatively small system could require many more batteries than are necessary for storage just to achieve the specified voltage level. Remember, too, that the same basic electrical laws of series and parallel connections apply to the photovoltaic modules as well. Choosing a high DC voltage for a small system could thus lead to the purchase of many more modules than the load actually requires just to achieve the specified array voltage. This situation, when taken in the extreme, would result in a rather unwieldy and quite uneconomical PV system.

In some cases, DC equipment is serviced directly from the photovoltaic system while AC loads are being served by the inverter. When this occurs, the system's DC operating voltage may be determined by the input voltage requirements of the specific DC-powered equipment being served. When there is a choice in operating voltages, use the general relationships expressed in table 5-2 as a guide. To select the optimum voltage, find the voltage category whose other parameters meet or exceed those of your system.

For example, a system designed to deliver 3.5 kilowatt-hours per day to a 1,600-watt AC load in an area with an average of 4 peak sun hours per day would require a photovoltaic array of roughly 1 peak kilowatt (kW_p). The recommended DC voltage for this system would be 24 volts. If, instead, the AC load requirements were 2,400 watts or there were only 3 peak sun hours available (increasing the array to about 1.5 kW_p), the general parameters established for the 24-volt system would be exceeded and 48 volts would be recommended. This is not to imply that you could not power the second system described with 24 volts. It's a matter

Table 5-2. DC Input Voltage Guidelines for Stand-Alone Inverters

PV Array Size (kW_p)	AC Inverter Load (kW)	Total Load (kWh/day)	System's DC Voltage (V)
Less than 0.4	Less than 1	Less than 1.5	12
0.4–1.0	2.5 or less	Less than 5.0	24
1.0–2.5	5.0 or less	5.0–12.0	48
More than 2.5	More than 5.0	12.0–25.0	120

of trade-offs in cost and efficiency. The guidelines are not hard and fast, and there will always be exceptions.

In broadest terms, SA inverters can be divided into three general categories based on their output *power capacity*. The first category comprises the "superlight" inverters with outputs between 25 and about 300 watts. These are primarily designed for use with an individual appliance on a periodic basis and are turned on and off manually along with that appliance. They are a useful option for photovoltaic users who have comfortably come to terms with direct current for all but one or two functions.

The second group consists of "medium-duty" inverters, those ranging in capacity from roughly 500 to 1,000 or 1,500 watts. Like the superlight inverters, these are generally used to run specific appliances within their wattage rating. They may also be used to serve two or more small loads in larger systems. When we move into this class of inverter, we begin to find automatic *load-demand switching* as an option on a few of the more sophisticated, higher-wattage models. Load-demand switching is accomplished by an electronic circuit internal to the inverter that keeps the unit in a low-power-consumption "idling" mode until it senses that a load requiring AC power has been turned on. At that point the inverter is switched on to full power.

Finally, there are the bigger "heavy-duty" inverters, which put out from 2.5 to 15 kilowatts of power at 120/240 VAC and are equipped to handle sizable surges. For any total load demand exceeding 1.5 kilowatt-hour per day, a heavy-duty inverter is the hardware of choice. These big inverters can also benefit greatly from load-demand switching, since standby losses are roughly proportional to inverter size and the large appliances and other loads for which they are designed are often not in continuous operation.

In many applications, an argument can be made for a dual-inverter system. In such a system a heavy-duty unit is employed only with compatibly heavy loads, while a small to medium-size high-efficiency inverter handles small loads. The object is to avoid operating any inverter for too long at too low a percentage of capacity. If the system is large and complex enough to warrant full-time service from a big inverter, the photovoltaic array will need to be sized large enough to provide for the unavoidable idling losses as part of the load. An additional reason for wishing to have a second inverter in the system might be to power one or two small loads that require a high-quality 60-hertz sine wave.

Selecting a Stand-Alone Inverter

Freedom from the special requirements of utility interconnection, such as fidelity to the utility-generated wave shape, makes the internal electronic logic and power conversion circuitry of an SA inverter different from that of a UI unit. However, many of the basic output requirements are common to the two inverter types.

Even though the SA inverter does not have to match the signal on the grid, wave shape remains an important consideration in its selection. The available output wave form options vary from a basic square wave to a quasi-square wave to a modified quasi-square wave to a near-perfect sine wave. The illustrations of these wave shapes shown in figure 5-3 are taken from SA inverters currently on the market.

The basic square wave is typical of nearly all of the superlight and many of the medium-duty SA inverters. The square wave is the least desirable output wave form because by definition it contains nearly 50 percent distortion and its peak voltage is much lower than that of a sine wave. In addition, many square-wave inverters lack even basic voltage and frequency regulation. This can result in poor performance of most motors and nearly all electronics with internal rectifiers.

The quasi-square wave is a somewhat less discordant compromise between the ideal sine wave and the basic square wave. Quasi-square-wave units can achieve the desired peak-to-RMS voltage ratio of a sine wave and their harmonics are lower than those of square-wave units, as shown in figure 5-3. Nearly all SA inverters over 1.5 peak kilowatts produce a quasi-square-wave output. These units often have good surge characteristics and can power most common loads with reasonable efficiency. The modified quasi-square wave is an improvement over the quasi-square wave because it further reduces harmonic-generated RFI interference and thus more closely approaches the sine wave ideal.

For the operation of synchronous and induction motors as well as many electronics loads, the sine wave is preferable to the quasi-square wave. Unfortunately, in the past, the cost of the pure-sine-wave inverter was higher and its conversion efficiency somewhat lower than that of the quasi-square-wave inverter. Also, it often lacked the ability to deliver and withstand heavy load surges.

It is expected that the use of new, state-of-the-art semiconductors and the higher-frequency switching techniques that they make possible, together with higher average DC input voltages, will allow SA inverters to catch up with the impressive recent advances in their UI counterparts.

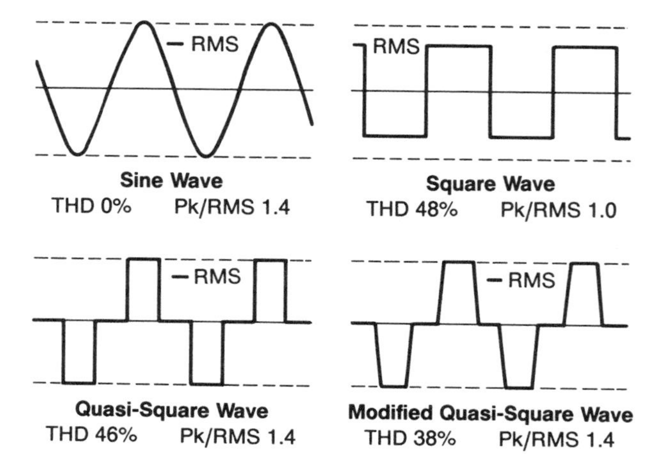

FIGURE 5-3
Output wave forms for four stand-alone inverters.

These improvements will soon make high-efficiency SA inverters that deliver a near-perfect sine wave output and have good surge-handling capability widely available at reasonable cost.

When you select an SA inverter, it is important to understand the wave-form quality requirements of the load you wish to serve. A quasi-square-wave inverter will be adequate for most loads; however, certain sensitive electronics loads such as personal computers will deliver optimum operation only with a sine wave.

Frequency stability is also important for the SA inverter, which must develop and maintain its own 60-cycle AC frequency. The amount of frequency control varies widely among SA units. Some of the poorer-quality models have little or no provision for it, while most sine-wave units have frequency deviations of plus or minus 0.005, and units with plus or minus 0.0001 are available. The majority of quasi-square-wave units fall somewhere in between in quality and design. The better units offer crystal control, which delivers very good frequency stability. Frequency control is important because it can affect certain electronics and motor loads such as phonograph turntables, clocks, VCRs, and tape decks.

Efficiency of operation is another factor that should influence the selection process. All inverters expend a certain amount of electricity on their own operation, independent of the current they supply to the load. Basically, the SA inverter experiences two types of internal losses. One is the *conversion loss* that occurs during operation when the inverter is delivering power to a load. The other is the *standby* or *idling loss,* which occurs when the inverter is waiting in the "ready mode" for a load to require power. The average heavy-duty SA inverter requires up to 4 amps of current (at its rated voltage) to operate in ready mode waiting for a load, while state-of-the-art units require less than 0.1 amp.

The full-load efficiency of an SA inverter will depend on its design. It can range from 50 to 60 percent for a small, 12-VDC superlight unit of the type you plug into your car's cigarette lighter to about 95 percent for a well-designed, high-quality unit. At an efficiency of 90 percent, a 5-kilowatt (nominal) inverter will require 5.5 kilowatts of DC input power to deliver its full-rated 5 kilowatts of AC power output. As the load is decreased to, say, 500 watts, the conversion efficiency of many average units will drop—perhaps to 50 percent or even as low as 40 percent.

The ideal inverter would have a *flat efficiency curve.* That is, the inverter conversion efficiency would be maintained at a high value throughout the entire operating range from less than one-eighth to full power. The relative "flatness" of the inverter's efficiency curve is a good indicator of the quality of the unit's design and translates directly into many kilowatt-hours of additional power delivered from the system.

***Flat efficiency curve*—efficiency maintained over the entire operating range of a piece of equipment such as an inverter**

Photovoltaic system designers must always keep in mind that compensation for an inefficient inverter can be made only by enlarging the PV array and batteries or by scaling down loads. When comparing the efficiency of various units, remember that some manufacturers have an overly optimistic view of their equipment's performance. If possible, obtain test data on the inverter from an independent third party and also speak with someone who has actually used the equipment under consideration in a similar application.

Another factor to consider in your inverter selection is *load-demand switching*. One good way of minimizing inverter standby losses is to equip the machine with an automatic load-sensing circuit that causes it to activate and start converting direct current to alternating current only when there is an immediate load demand. Nearly all higher-power SA inverters now come equipped with load-demand start as a standard feature, while others offer it as an option.

Load-demand switching—an automatic load-sensing circuit used to activate an inverter upon load demand, allowing it to remain off when not needed

Automatic load-demand starting enables the inverter to automatically switch itself on or off in the presence or absence of a load, eliminating the majority of the losses associated with idling during no-load periods. This is an especially important feature since an SA inverter used for anything but a dedicated (fixed) load is likely to see service for only relatively short periods each day. Without automatic load demand, an inverter with a high standby loss will expend a lot of electricity just idly waiting for a load to come on. Of course, if the use of an appliance or piece of machinery can be planned, the manual inverter on/off switch will work just as well—but *only* if it is turned off following each use.

One problem with many automatic load-demand circuits of early design is that they have a high starting threshold and typically require about 50 (or more) watts of connected load to activate the inverter. Some small appliances demand so little current that by themselves they are not capable of reaching the demand threshold and calling the machine into service. The answer to this problem is to either use the manual override switch or group small loads together—by making coffee and listening to the radio while you are shaving, for example. Today's higher-quality units have effectively solved this problem.

Another problem with the early-design load-demand starting circuits is that there can be a delay of up to several seconds from the time the load is turned on until the inverter's load demand responds and the load is energized. This can present some serious problems for people who are not familiar with or constantly mindful of this idiosyncrasy. Someone using power tools, for example, might find this particularly bothersome. But state-of-the-art units are always in the ready mode and able to power a load with no minimum starting threshold and no time delay.

The next characteristic that we should look for in an SA inverter is its ability to handle load surges adequately in accordance with the demands of the loads it is to serve. A load surge, as has already been mentioned, is the phenomenon associated with the start-up of certain appliances. It is the major contributing factor to high peak-minute and peak-second loads. Certain loads, such as large motors, demand a high inrush current to start. For example, an inductive motor rated at 1 horsepower (HP) will normally require between 1 and 1.5 kilowatts of power to operate (depending on motor efficiency), but may require 5 kilowatts or more of inrush current just to start. This is especially true of a motor such as a well pump or a refrigeration compressor that must start under full load.

Table 5-3. Power Requirements for Starting and Running AC Electric Motors

HP	kVA*	kW*	Starting Power Factor (%)	Locked Rotor (starting) kVA (per NEMA code letter)									
				A	B	C	D	E	F	G	H	J	K
0.12	0.40	0.25	80	00.38	00.44	00.49	00.55	00.62	00.69	00.77	00.87	00.98	01.10
0.25	0.65	0.40	80	00.80	00.91	01.02	01.15	01.30	01.43	01.61	01.82	02.05	02.30
0.33	0.72	0.45	80	01.05	01.20	01.35	01.52	01.70	01.90	02.13	02.40	02.70	03.04
0.50	0.97	0.60	80	01.60	01.81	02.04	02.30	02.60	02.87	03.22	03.63	04.08	04.60
0.75	1.37	0.85	85	02.39	02.72	03.06	03.45	03.85	04.30	04.83	05.44	06.12	06.90
1.00	1.60	0.99	80	03.20	03.60	04.10	04.60	05.10	05.70	06.40	07.20	08.20	09.20
1.50	2.20	1.45	80	04.80	05.50	06.00	06.90	07.70	08.60	09.70	10.90	12.00	14.00
2.00	2.60	1.80	90	06.40	07.30	08.20	09.30	10.30	11.50	13.00	14.60	16.40	18.50
3.00	3.70	2.80	90	09.60	10.90	12.00	13.90	15.50	17.00	19.40	21.90	24.60	27.70
5.00	5.40	4.80	90	16.00	18.00	20.40	23.00	25.80	28.80	32.40	36.50	41.00	46.00
7.50	8.40	6.50	90	24.00	27.50	30.50	34.50	39.00	43.50	48.50	55.00	61.50	69.50

*Operating at 100% load.

The starting surge requirements of induction motors can be determined from the manufacturer's information or the data on the motor's nameplate. When you know both the horsepower rating and the National Electrical Manufacturers Association (NEMA) letter code, you can determine the inrush power in kilowatts or kilovolt-amps (kVA) required for starting the motor by using table 5-3.

Kilovolt-amps (kVA)—electrical units similar to kilowatts that take power factor into account when multiplying volts times amps

The surge capabilities of different inverters are rated by their manufacturers in relation to and as a function of their *power capacity*, or the *maximum continuous load* that they are capable of handling. They are also rated in terms of the size (in horsepower) of the induction motors that they are capable of starting *one at a time*. Comparative ratings for a selection of typical SA inverters are presented in table 5-4. These ratings change as the technology advances, and relative prices should reflect an inverter's constant power output, its versatility in surge capacity, and its overall efficiency.

In a UI photovoltaic system, the utility grid itself will provide current to handle all load surges, thus freeing the line-tied inverter from this responsibility. When you are on your own, though, handling peaks and surges becomes a matter of having an appropriately powerful inverter and perhaps a working load-management plan as well.

Table 5-4. Typical Stand-Alone Inverter Power and Surge Capacity Ratings

Inverter Rating (kW)	DC Input Voltage (V)	Continuous Load (kW)	Surge Load (kW)	Maximum Induction Motor Load (HP)
2.0	24	2.0	10.0	¾
2.5	48	2.5	10.0	1
5.0	48	4.0	20.0	2
6.0	120	5.0	24.0	3
12.0	120	10.0	48.0	5

In larger systems or critical-reliability applications, a microprocessor-based load-management device is often employed to provide dependable, automatic load management. In smaller applications, such as remote residences, this will often be accomplished by simply practicing discretion in the simultaneous operation of household equipment. Stated in concrete terms: don't expect to use your radial arm saw while the washing machine, the well pump, and the (gas) dryer are all running, your wife is using the blow dryer, and the kids are still in the Jacuzzi. Life with a small SA system isn't exactly the same as living on the utility grid, especially if your system has only a 1.5-kilowatt inverter.

We now come to the matter of voltage stability. Stand-alone inverters, unlike their utility-interactive counterparts, must establish *and* maintain or regulate their own AC output voltage. This is a somewhat difficult task, since in addition to constantly varying load demand, the system's DC input voltage is continually drifting up and down in relation to the battery's state of charge. The small, superlight inverters offer little or no voltage stability, while the quality sine-wave units deliver within 1 to 2 percent of their prescribed AC output voltage. The quasi-square-wave units all fall somewhere in between, with the average "good-quality" units able to deliver within 5 to 10 percent of the desired goal, and state-of-the-art units matching the regulation specifications of the sine-wave inverter. Most common AC loads should operate quite adequately with voltage variations of between 5 and 10 percent. In fact, utility regulation of power on the grid is often not much better than this, and sometimes it's worse.

Every method of voltage regulation has a set of limits within which it operates. Most quality SA inverters now feature high- and low-voltage cutouts that monitor the level of the DC input voltage. If input voltage excursions take the regulation circuitry to the limits of its working voltage "window," the inverter shuts down. This is an important feature, one that protects not only the inverter but the loads and the battery bank as well.

The same power and signal quality requirements imposed by utility interface in a line-tied photovoltaic system are shared by many of the loads likely to be connected to an SA system. Some appliances—most notably record turntables, VCRs, tape recorders, clocks, and other devices with synchronous induction motors—require a clean and constant 60-hertz frequency for optimum operation. Harmonic distortion above 15 or 20 percent can cause these induction motors to overheat, and a voltage variation in excess of 10 percent above or below 120 volts may end their useful life prematurely.

Harmonics are present in the basic non-sine-wave output wave forms shown in figure 5-3. They can also occur when spurious signals are mixed with the inverter's output to distort its wave form. These signals can be caused by reactive loads and other external sources; however, they most often come from the switching of the power semiconductors within the SA inverter itself. Harmonics generated within the inverter can be filtered out. However, since filter circuits add to the cost and internal losses of the unit, it is a far more desirable strategy to design the inverter's basic electrical topology to deliver a clean wave form free of switching spikes and other harmonics so as not to require extensive filtering.

The superlight inverter's total harmonic distortion can vary wildly depending on operating conditions, since its basic square wave has a 48 percent THD to start with and it has little or no output filtering. The sine-wave units can deliver AC power with a THD of 5 percent or less and the average quasi-square-wave units deliver somewhere between 20 and 40 percent THD, a good part of which is present in their basic wave shape.

Maintaining a good power factor is also important in an SA system and depends upon the selection of both the inverter *and* the load equipment, since both can be sources of reactive power. Most quality, heavy-duty SA inverters deliver an acceptable power factor of 0.7 or better under most load conditions. However, some units have trouble maintaining a suitable power factor if they are operating multiple loads that exhibit different inductive or capacitive characteristics. This can result in surges of reactive power flowing through the distribution system between the inverter and the inductive load. In extreme cases this can cause the lights in your house to blink off and on marquee-style when your washing machine is running.

EMI and RFI generated by an SA inverter can be troublesome intrusions into the world of home electronics. But fortunately, their elimination is not exceedingly difficult. Filters and grounded shielding can be placed on the offending equipment that produces unwanted signals. Or filters can be placed in the line at the power cord of the affected electronics gear. However, by far the best solution to this nuisance is to choose an inverter whose basic internal circuitry is designed so that it does not produce any interference.

Stand-alone inverters, like their line-tied counterparts, produce audible noise. The level of noise created will vary depending on the inverter's basic circuit design and the type and amount of load placed on it. Before choosing an inverter, always check to see how much noise it generates. When installing it, locate it away from noise-sensitive areas.

Last but not least on our list of factors to consider when selecting an SA inverter are *reliability* and *maintainability*. The inverter is by far the

Table 5-5.
Comparison of Stand-Alone Inverters

Inverter	DC Input Voltage (V)	Continuous Load (kW)	Overall Efficiency (%)	Wave Form Harmonics
Heart	24	2.5	96	Good
Best	24	2.0	82	Fair
Nova	24	2.5	90	Excellent

most complex, vulnerable, and unpredictable component in a photovoltaic power system. Therefore, it is important to select a unit that is going to last a long time and is easy to repair if it ever breaks down. But how does one assess these qualities in a particular inverter? The best way is to check on its field record. If possible, talk with someone who has used the same equipment in a similar application.

Inverters are heavy and bulky items. This makes them costly and time-consuming to send across-country for diagnosis and repair. Look for a unit that features modular design and construction, since this will allow easy replacement of component parts. Most inverter failures can be traced to the main logic board or the power semiconductors. If these parts can easily be replaced, you can be back in operation after minimal downtime. So look for an inverter that can be repaired in this way, has a good record of field-proven performance, and is represented by a reliable local professional who fully understands how it works and how to fix it when it does not.

Obviously, the inverter is a very important piece of equipment in both SA and UI photovoltaic systems. It should therefore be a high-quality piece of equipment, one that is carefully selected to suit the needs of the particular system. The PV system designer should be intimately familiar with the manufacturer's design and performance specifications as well as with the installation and operation procedures for every unit being considered. Table 5-5 provides a comparison of some better-known stand-alone inverter models.

Backup Generators

A fossil-fuel-fired backup generator may quite logically be considered as one of the balance-of-systems components of an SA photovoltaic installation. The reason is simple: in order to cover relatively infrequent instances of high load demand and/or low array output due to prolonged adverse weather conditions, a PV system that lacks a backup generator

Surge Capacity (%)	Voltage Regulation (%)	Frequency Stability (%)	Retail Cost ($ approx.)
400	2	0.10	2,000
500	5	0.15	2,200
200	1	0.15	3,200

must be substantially oversized in terms of both array capacity and battery storage. In effect, this amounts to devoting a lot of money and space to insurance capacity that will not usually be needed. The generator solves the problem by standing ready to boost the battery charge should it fall low or to help the system through periods of a typically high demand. A backup generator thus makes sense for any year-round stand-alone residential PV installation with even a modest assortment of loads, especially in a climate in which daily sunshine cannot be taken for granted.

A few numbers can help bring home the point. Even at a module price of $8 per peak watt, an array sufficient to service a daily load of 5 kilowatt-hours would cost $15,000 (1,875 W_p output with 3.0 hours of average peak sun). Batteries sufficient to provide ten days (50 kWh) of storage—a reasonable insurance against bad weather—would cost another $13,500 at today's prices. However, if the array were sized to supply a battery bank from which only two days' storage was required, the module cost could be brought down to perhaps $12,000 (for 1,500 W_p based on a more optimistic design value of 3.5 average peak sun hours) and the battery cost reduced to $3,000 or less.

The difference between these comparative array/battery investments of $28,500 and $15,000 is the inclusion, in the latter system, of a generator (or "gen-set") and battery charger with a combined additional cost of about $4,500. The net savings, then, is $9,000, with the added benefit of increased system reliability and flexibility through diversification. The initial cost of this hypothetical system could be reduced even more by further reducing the array size, if the owner were willing to rely on the generator to a greater extent.

For example, a 500-peak-watt array could be installed initially, with provision made for future addition of photovoltaic modules to the array as finances allowed. Gradual reduction of generator operation would coincide with expansion of the array size. This approach would reduce the initial cost of the system by another $8,000 and still meet the load requirements, while providing many of the benefits of a hybrid system and allowing array capacity to be increased as desired.

When designing a photovoltaic/generator hybrid system, the gen-set should be sized so that its *continuous duty rating* equals or exceeds the total of the maximum power draw of the system's battery charger *plus* any AC loads in the system that are to be served during generator operation, with at least a 10 percent margin of extra capacity remaining. At the same time, be careful not to oversize the generator. The variable nature of residential loads almost guarantees that there will not be a great deal of load requirement at the specific time the generator is called upon. If the site of your installation is at an elevation significantly above sea level, make sure you de-rate the generator as needed to allow for the effect of high elevation on the engine horsepower and the resulting reduction in generator power output.

Generators have been around a long time and have evolved into many sizes and types. There are three fuel choices: gasoline, propane (natural gas), and diesel. Gasoline and propane generally fuel only the smaller to mid-size residential gen-sets (up to 8 or 10 kW), with diesel taking over as the more commonplace option in the heavy-duty, mid-size to large range. There are, of course, gasoline and propane generators sized large enough to take care of any residential loads. Some units are available with optional dual-fuel carburetion so they can accept gasoline or propane.

Propane is favored over gasoline as a generator fuel, although diesel-fired units are the most durable.

Propane is generally favored over gasoline because propane units are quieter, run cleaner, and offer a modest reduction in maintenance requirements. This choice is especially easy if propane is already being used on-site for other household functions such as cooking, refrigeration, or drying clothes. Diesel, on the other hand, is a handy fuel to have on hand if you own a diesel automobile. But the best reason for going with a diesel machine is its general superiority in terms of overall ruggedness. All other things being equal, diesels will outlast propane units, and they are sure to outlast gasoline-fueled units. However, diesels start harder in cold weather, run louder, and produce a more undesirable exhaust than do propane units.

Whatever the choice of fuel, it is best to choose a unit with a governed operating speed of 1,800 RPM rather than 3,600 RPM for all but the smallest, light-duty applications. The slow-turning gen-sets last longer, run quieter, and need less frequent maintenance. Look for heavy-duty equipment with pressurized, pumped lubrication and a replaceable oil filter. Liquid cooling of the engine block is preferred, as this runs quieter and also provides more uniform cooling, reducing stress on internal parts and prolonging engine life.

Don't try to skimp on a generator. Avoid the units with modified lawn mower engines for all but the smallest, occasional-duty applications. It's better to put more money into your initial investment and have fewer worries about reliability, expensive repair, and subsequent replacement.

Since the gen-set will be your backup source of power, it must have the ability to start on demand. All but the smallest gen-sets have electric starting. If automatic generator start-up is desired, an interface will be required with the low-voltage sensing circuit in the system's voltage regulator, which determines the point at which the battery bank has been discharged to its recommended depth of discharge. The preferred interface will automatically start the generator and then transfer the system's AC loads from the inverter to the generator when the battery bank reaches this specified level of discharge. Generator operation will continue to charge the battery bank via the battery charger until a certain preset level of charge is reached and/or until battery charging resumes from the photovoltaic array. At that point, the generator will be shut down and the system will return to normal operation with the inverter supplying the AC loads.

Another feature to look for in a generator is an automatic, timed "exercise" function. A preset timer turns the machine on for a period long enough to circulate oil and coolant and keep the seals lubricated. Too long a period of inactivity can be harmful to an auxiliary generator or any other internal-combustion engine. A sophisticated gen-set will have a built-in override for this function, by which the timer is reset forward if the engine has recently been run for battery charging. Automatic shutoff, with warning indicators, should also be provided in the event of engine overheating or low oil pressure.

Generators can be equipped with many special features, such as preset timers and heat-recovery devices.

One option offered by some manufacturers of liquid-cooled generators is a water circulation system that allows the recovery of otherwise wasted heat. Kohler Company manufactures gen-sets in five sizes, configured to circulate engine coolant through space-heating convectors or other external equipment. Heat recovery from the hot exhaust gas stream is also frequently done with the aid of special liquid heat exchangers that are installed in the exhaust gas piping before the muffler. The recovered heat can

PHOTO 5-4

This 10-kilowatt engine generator set manufactured by the Kohler Company features automatic starting, exercising, and safety controls.

be delivered to a storage tank for space heating and domestic hot water heating.

Generator heat recovery makes the most sense, however, when the gen-set is to be operated frequently and for long periods of time. This situation does not usually apply when the photovoltaic array is the primary source of power and the engine spends most of its time in a standby mode with less than 50 hours of operation per year. Ironically, for cold climate operation, this heat transfer is very often done in reverse, with the PV power system supplying heat to the gen-set via an electric crankcase heating element to ensure reliable winter gen-set operation, especially for diesel units installed in unheated areas.

Other Photovoltaic System Controls

A number of special options can be included in a photovoltaic array/battery/auxiliary generator system to assist its interface with electrical loads. One example, mentioned earlier, is automatic transfer switching of loads from the inverter to the generator. In more sophisticated systems, an automatic control will start the generator at a preset low battery voltage threshold. Then, after the generator is up to speed and has reached operating temperature, the switch will transfer the load from the inverter, allowing the generator's AC power to directly serve the system's AC loads. The system's DC loads will be served by the DC output from the system's battery charger, which is in operation when the generator is running.

There is also an assortment of individual load-management devices that can be installed to serve specific, heavy-demand appliances. Their use results in the cutting of overall demand, the shaving of peak loads, and a lightening of the tasks of the array, batteries, and auxiliary generator. These differ from the microprocessor load-management units discussed in chapter 6 in that they do not prioritize and shut off one or more of the selection of loads but instead monitor and efficiently limit the demand of the individual appliances to which they are dedicated.

One such device is the Frige-Mate, manufactured by Computerized Technologies, Inc. The unit, which sells for under $100 and accommodates loads of up to 10 amps and 1 horsepower at 115 VAC, makes six calculations per second of appliance current requirements relative to load conditions. In effect, it gives the appliance only the current that it needs without endangering motors; if voltage input is too low, the device shuts off the load until the required level is again reached. Torque and RPM are not affected, and motor life is said to be extended by 30 percent due to cooler operation. The company also makes a Motor Genie model, in 115- and 220-VAC versions, with 5-horsepower/30-amp capacity. This $160 unit is recommended for loads such as central and large room air conditioners, swimming pool pumps, and Jacuzzi motors, while the Frige-Mate is sized to serve washers, refrigerators, freezers, and small air conditioners.

Both products are guaranteed to cut electricity use in the appliances served by at least 20 percent.

Another load-reduction product is the Alden-Mason Company's Energy Sensor Energy Management Controller, designed to function with central air conditioning or heating systems. It monitors the interface between the thermostat and the furnace or air conditioning control system and constantly searches for the shortest possible running time on the basis of how long the unit had to run to satisfy the thermostat on previous cycles. With air conditioning, the controller provides for a 2-minute fan run after the compressor shuts down to take advantage of cooling capacity remaining in the coil. The manufacturer suggests an average payback period of two years for the $700 to $1,200 product, based on the purchase of a package that includes a custom thermostat.

Monitoring Devices

The remainder of the components in most photovoltaic systems fall largely into the category of monitoring devices not strictly essential to system operation. In cases where a grid connection exists, utility companies will require the installation of a meter that will accurately record the number of kilowatt-hours drawn from and sold to the utility. Some utilities will accept a single, nonratcheted meter that will run forward when electricity is being sold to the homeowner and backward when the array output exceeds home load demand and power can be sold back to the utility. On

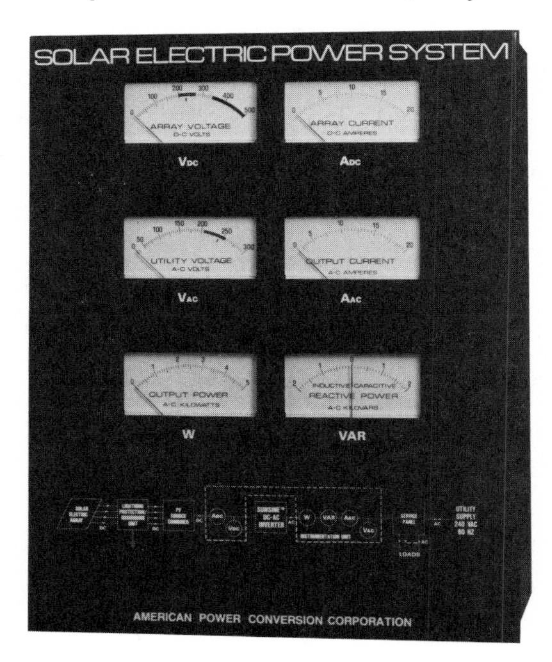

PHOTO 5-5

This monitoring package for utility-interactive PV systems by American Power Conversion Corporation measures array voltage and current, utility voltage, inverter output current and power, and reactive power.

systems designed for loads larger than a single residence, however, most utilities will require two separate ratcheted kilowatt-hour meters—one to record incoming power (purchases) and the other to record excess output fed into the grid (sales). In some installations, this single or double kilowatt-hour metering arrangement may be accompanied by a third meter that keeps track of the PV system's total DC output.

Aside from these basic kilowatt-hour meters, there is an assortment of measuring hardware that will appeal to photovoltaic users drawn to technical detail. The auxiliary devices include pyranometers, which measure both diffuse and direct solar radiation at a particular site, and a variety of meters that can measure DC voltage and current, AC voltage current and frequency, DC amp-hours and even VARs, which are units of reactive power. Standard instrumentation packages are available for a variety of PV applications, such as those manufactured by American Power Conversion Corporation.

This chapter was primarily concerned with what are (after the array, batteries, and battery charger) the most important components in a stand-alone photovoltaic system—inverters and backup generator sets. In the following chapter, we will look at the numerous and diverse factors that combine to create the load demand in the modern home, no matter what power system it uses.

CHAPTER SIX
DETERMINING YOUR ELECTRICITY REQUIREMENTS

It might be instructive to remind ourselves that we are approaching the end of what has been only the first full century in which mankind has used electricity. Because electrical service is available nearly everywhere in the developed world, and because (at least on our side of the wall outlet), it seems like such a simple and dependable commodity, it's hard to remember that we did not always have it at our disposal. But less than 200 years ago, electricity was just a laboratory curiosity; it fascinated far-thinking scientists like Benjamin Franklin but did no practical work in the world.

Slightly less than 150 years ago, Samuel Morse first applied electricity to the task of sending coded messages. And so the age of rapid communications was born. In the 1880s, Thomas Edison began to explore many more uses for electrical current. His chief accomplishment of that era was the incandescent lamp, but even this great stride forward was at first hindered by the need to generate electrical power at or very near to the point of use. Edison's Pearl Street generating station in lower Manhattan was a crude, direct-current affair capable only of lighting an area of just a few city blocks.

It wasn't until Edison lost his celebrated "battle of the currents" with alternating-current advocate George Westinghouse that electricity was practically and efficiently sent over long distances. Still, for the next 50 years or more, electric lights and appliances remained largely an urban phenomenon. In the 1930s, the electric distribution grid began to reach out into the countryside under President Roosevelt's Rural Electrification Program. However, in some areas, it reached very slowly. Several towns in Vermont less than 300 miles from Boston had no electricity before 1961, and some remote Western ranches still must generate their own power.

But by the end of World War II, the great centralized American power grid was largely in place. Economists began to talk about a direct relationship between growth and electrical consumption, a notion still alive even though it has been largely disproved by the success of conservation programs that have improved economic expansion. For some time now Americans have been accustomed to thinking in terms of a future in which the available amount of electricity is virtually limitless. A study of residential photovoltaic (PV) system design can help us focus more clearly on the relationship between the desire for electrical power and the cost of producing it.

Electricity: The Premium Form of Energy

Electricity is our premium form of energy; yet, to most people, it is also an abstraction, something taken for granted but not well understood. Essentially, there are two ways in which it appears tangible to most of us. First, it provides heat for our bodies and light and power for our appliances. Second, it appears in the "Balance Due" column of our utility bills. If it were delivered in a truck and stored like home heating oil in a tank whose gauge read "empty," "half," and "full," perhaps electricity might be easier to conceptualize and thus to conserve. But few of us think in terms of electrons flowing through our dishwashers and floor lamps like so many gallons of oil, and the concepts of amperes, watts, and kilowatt-hours have until recently remained essentially the specialized concerns of electrical engineers.

In order to seriously address the challenges of energy production, conservation, and self-sufficiency, we need to begin to appreciate electricity as a *commodity*—similar, in this respect, to petroleum, coal, or natural gas. This is important for two reasons. First, because at present the availability of utility-generated electricity is directly correlated to our finite stores of the more tangible fossil fuel commodities and to the extremely problematic business of nuclear power plants. Second, because even apparently limitless sources of electricity, such as solar-powered photovoltaics, must be put to work with some measure of prudence if they are to be practical and cost-effective. There may be an inexhaustible supply of sunlight, but there are not unlimited amounts of roof space or money.

It is a far better investment to incorporate energy-conscious design features along with state-of-the-art, high-efficiency appliances and equipment into homes and commercial structures than it is to build a bigger electrical generator to supply a load made unnecessarily large by inefficiency. This is true *no matter what kind of generator you build.* With efficient design, you can deliver the same level of comfort and convenience with a higher level of reliability, security, and sustainability at a lower cost.

To appreciate electricity as a commodity, we must have a clear sense of how it is used and how it can be conserved. This means understanding the quality and amounts in which it must be supplied to our houses. Such knowledge is prerequisite to the sizing and design of properly proportioned and cost-effective PV systems.

Energy Consumption in the Home

The concepts of voltage, amperage, and wattage have already been used frequently as a means of defining the power output of a PV system. A more detailed explanation of these terms, as well as of kilowatt-hours, can aid our understanding of electricity as a measurable commodity.

Voltage is the term applied to the electromotive potential or "electrical pressure" in a generating device, battery, or electrical circuit. The prevailing line voltage in the United States—that is, the voltage supplied by utilities to residential and small to medium-sized commercial customers—is 115/230 volts of alternating current (VAC). (Although the designations of 110/220 VAC or 120/240 VAC are sometimes applied to line voltage or to home appliances, they are for all practical purposes synonymous with the 115/230 VAC standard.)

As we saw in our survey of system options in chapter 2, small-scale, independent photovoltaic systems can be designed to run on only direct current using as little as 12 volts, providing that only DC appliances of the proper voltage are used. But when a DC-only system is contemplated, loads must be kept low and the resultant trade-offs on system size and efficiency must be considered.

Consequently, we can assume that the end product of almost all residential PV systems must be compatible with the 115/230 VAC norm. The 110 to 120 VAC requirements of appliances, equipment, and lighting fixtures commonly sold in the United States will likely dictate adherence to this standard even in PV-powered houses not linked to the utility grid.

Voltage, as noted, is a measure of the pressure or potential, as opposed to the flow, of electrons. The flow is called *current*, the measurement of which is expressed in amperes or amps. One amp, in turn, is equal to 1 *coulomb* (6.24×10^{18} electrons) passing a given point in 1 second. That is 6.24 quintillion electrons galloping past an arbitrary checkpoint with each tick of the clock.

Coulomb—**the quantity of electricity transferred by a current of 1 ampere in the space of 1 second**

Voltage and amperage are the two basic terms used in measuring electricity as an abstract force. Another term, the *ohm*, refers to electrical friction or resistance. Ohm's law tells us that the current in amps that will flow in a given circuit will equal the applied electromotive force in volts divided by the resistance (R) in ohms ($A = V \div R$). Stated another way, the voltage drop across a load is equal to the current through the load in amperage and the resistance of the load in ohms.

When we wish to speak in terms of the practical application of electromotive potential and current, of the work to be accomplished by electricity, the concept of wattage comes into play. Watts are units of power or the rate of flow of energy. Wattage is the product of voltage times amperage: 1 ampere (A) flowing with the electromotive potential of 1 volt (V) produces 1 watt (W), which also represents one joule (J) of energy. A typical two-slice toaster, using 115-VAC household current at 9 amps, is rated at slightly over 1,000 watts ($115 \times 9 = 1{,}035$), or 1 kilowatt (kW).

The kilowatt is the basis for the unit of measurement of residential electrical consumption, namely the kilowatt-hour (kWh). One kilowatt-hour is simply 1,000 watts of electricity used for the period of 1 hour. Utility company meters record electrical use in kilowatt-hours and base their rates upon this unit of measure. So when you want to know how much electricity individual appliances in your home consume daily, monthly, and yearly, you need to calculate this in terms of kilowatt-hours.

For instance, if a 1-kilowatt toaster is used an average of 6 minutes each day, its daily electrical consumption is 0.1 kilowatt-hour, which amounts to an annual total of 36.5 kWh. Knowing the potential kilowatt-hour demand of each load helps you see the value of selecting energy-efficient appliances. At the same time, knowing the anticipated (or proved) kilowatt-hour requirements of an entire house is essential in planning and sizing a PV system.

Where Does the Electricity Go?

Just as the "average" American citizen exists only in the minds of statisticians, so too we will search in vain for the American house exhibiting "average" energy-use patterns. Nevertheless, it is possible to estimate with reasonable accuracy the electricity consumption of various houses. The wattage requirements for household appliances are readily available. (Your local utility company is one possible source.) Since a house's load profile is an aggregate of the number and type of electrical devices in use and the amount of time each is in service, the calculation of projected kilowatt-hour use is a matter of simple arithmetic.

Looking at the kilowatt-hour reading on your electric bill every month will help you determine your total electricity consumption, but it does not answer the question of where the power is going. To depend solely on that figure for your energy profile would be like getting your monthly bank statement and having no itemization of checks written or like setting a home budget as a lump sum without categorical allotments. To manage money and spend it wisely, you have to know where it is going. The same is true of electricity.

One of the main reasons for the following exercise in tracing electricity use patterns is to cultivate an awareness of electrical energy as a commodity that can be budgeted and applied to specific purposes in desired amounts. This is crucial to any rational decision regarding domestic photovoltaics. Particularly when sizing a stand-alone (SA) photovoltaic system, it is crucial to know *how much* power is going to be needed and *when* it will be called for, so that you can establish an accurate "load profile" for your application.

House 1

To begin our exercise, let's imagine three different houses. House 1 is a moderate-sized three-bedroom residence with four inhabitants inclined toward relatively moderate use of electricity. It is not electrically heated (in fact, none of the three houses has electric resistance space heating) and has one room air conditioner that is used only on the hottest days. There is an electric range but no self-cleaning oven. The dishwasher is used only for large loads, once a day at most. Domestic hot water (DHW) is not heated electrically. The house contains an assortment of small electrical appliances, although they are used infrequently and do *not* include such items as dehumidifiers, trash compactors, and electric blankets. The refrigerator is a mid-size, manual-defrost model, and lighting is partially fluorescent. Finally, most of the appliances in use in this house are several years old and thus not as efficient as the state-of-the-art models discussed later in this chapter.

Table 6-1 profiles the monthly kilowatt-hour use for our hypothetical House 1, with monthly kilowatt-hour figures provided for each appliance and the total rounded off to the nearest tenth. You will notice that these figures are determined by multiplying the kilowatt rating for each appliance by the hours of use per average day and then multiplying that total by 30. Of course, not all appliances are used each day. In such cases (the air conditioner is one example), a daily average is taken based upon monthly use.

Table 6-1 represents electricity use in House 1 exclusive of lighting. If we add a moderate estimate of 75 kilowatt-hours per month (yearly average) for a combination of fluorescent and incandescent lighting fixtures, the monthly figures will read as follows: winter, 490.2 kilowatt-hours; summer, 487.2 kilowatt-hours; spring or fall, 401.2 kilowatt-hours.

House 2

Now picture a second, somewhat larger house in which four to six persons live in a way that demands a more extravagant use of electricity. In order to avoid repetition, let's forgo compiling an account of electricity consumption as in table 6-1. After all, a good part of the difference between the two houses' electrical consumption may be attributed to more frequent or prolonged use of the same appliances. In addition, the second house is more likely to contain a greater assortment of appliances, each of which by itself may not contribute significantly to the load but when taken together can appreciably boost electricity consumption. An electric blanket, for instance, consumes about 12.5 kilowatt-hours per month and an electric griddle about 3.7 kilowatt-hours. For the sake of this comparison, then, let's assume that the second house bears a load increase, related to the additional use of the basic appliances, of approximately 20 percent of the values in table 6-1 (before adding lighting). Adding an even 20 percent to our previous figures, we get subtotals of 498.2 kWh for winter, 494.6 kWh for summer, and 391.4 kWh for spring and fall.

To these figures, add the kilowatt-hour consumption of the additional major appliances listed in table 6-2. Note that the air conditioners are in operation only in the summer, which makes the subtotal for the summer considerably higher than that for spring or fall. The winter subtotal includes 17.8 kilowatt-hours more than the subtotals for spring and fall,

Table 6-1. Electrical Consumption in House 1

Appliance	Wattage (kW)	Hours Used/ Day		kWh/ Month
Blender	0.40	0.015		0.20
Clock	0.002	24.00		1.40
Clothes dryer	4.86	0.50		73.00
Electric shaver	0.015	0.08		0.04
Dishwasher	1.20	1.00		36.00
Garbage disposal	0.40	0.06		0.70
Heating & DHW				
Oil burner motor	0.25	8.00	(winter only)	60.00
Hot water circulation pump*	0.12	8.00	(winter only)	29.00
Iron	1.10	0.30	(current flows 50% of this time)	5.00
Kitchen exhaust fan	0.20	1.00		6.00
Mixer	0.15	0.10		0.50
Oven	2.60	0.50	(current flows 25% of this time)	9.80
Portable hair dryer	0.80	0.10		2.40
Radio-phonograph	0.10	3.00		9.00
Range				
1 large burner (high)	2.40	0.30		21.60
2 small burners (high)	1.30	0.40		31.20
Refrigerator-freezer (manual defrost)	0.32	24.00	(current flows 40% of this time)	92.20
Room air conditioner	0.86	10.00	(10 days/month; summer only)	86.00
Sewing machine	0.08	0.15		0.40
Television (solid-state color)	0.20	4.00		24.00
Toaster	1.00	0.10		3.00
Vacuum cleaner	0.65	0.12		2.30
Washing machine	0.50	0.50		7.50

Subtotals (kWh/mo):

Winter: 415.20 (heating system in operation)
Summer: 412.20 (air conditioner in operation)
Spring or fall: 326.20 (no heat or air conditioning)

*An oil-fired, hydronic, baseboard space-heating system is assumed here. If a hot-air system were used, the circulating fan would require about 80 kWh/mo.

Table 6-2. Electrical Consumption of Additional Major Appliances in House 2

Appliance	Wattage (kW)	Hours Used/ Day		kWh/ Month	
Freezer (14 cu ft; manual defrost)	0.44	24.00	(current flows 50% of this time)	158.00	
Oven (self-cleaning)	2.50	0.10		7.50	
Room air conditioners (2 additional)	0.86 ea.	10.00	(10 days/month; summer only)	172.00	total
Television (black & white)	0.06	2.00		3.60	
Television (color; additional)	0.20	2.00		12.00	
Water heater (80 gal)	3.00	24.00	(current flows 19% of this time)	410.00	

Subtotals (kWh/mo):

Winter: 608.90 (includes extra 17.80 for heat)
Summer: 763.10 (air conditioners in operation)
Spring or fall: 591.10 (no heat or air conditioning)

which represents a 20 percent increase over the amount of electricity required to run the heating system (oil burner motor and hot water circulating pump) in the first house. This brings our winter subtotal to 608.9 kWh per month.

Table 6-3 shows the sum of the two sets of subtotals for House 2.

We still have not accounted for increased use of electric lighting. In House 1, a lighting load of 75 kilowatt-hours per month was assumed. Given several additional rooms, less attention to conservation, and some high-energy-use items such as track lights and outdoor floodlights, this figure could easily double. Thus, the final monthly figures for House 2, with an additional 150 kilowatt-hours per month added for lighting, are as follows: winter, 1,257.1 kWh; summer, 1,407.7 kWh; spring and fall, 1,132.5 kWh.

House 3

Still using House 1 as a basis for comparison, imagine a more modest, energy-efficient residence in which electricity is expended conservatively in patterns far more in tune with the cost-efficient use of photovoltaics. This passive solar, superinsulated house achieves most of its energy savings by conserving heating fuel rather than electricity. Those savings will not be reflected in an energy profile that concerns itself exclu-

Table 6-3. Electrical Consumption in House 2

	Winter (kWh/mo)	Summer (kWh/mo)	Spring or Fall (kWh/mo)
Base	498.20*	494.60*	391.40*
Additional	608.90	763.10	591.10
Total	1,107.10	1,257.70	982.50

*kWh/month for House 1 plus 20 percent for increased use.

sively with kilowatt-hours. However, many of the same features of insulation and passive solar design that make heating fuel savings possible also contribute toward decreasing or eliminating the need for air conditioning. This is true in hot climates as well as in the more temperate zones.

Major savings in electricity consumption can be accomplished in House 3 by not using an air conditioner (86 kWh per month) and by cutting in half dependence on the clothes dryer and dishwasher (savings of 36 and 18 kWh per month, respectively). For the sake of this last example, assume that clothes are dried outdoors in good weather and that the dishwasher is used only when there is an especially large load. Further energy savings can be accomplished by lowering water temperature settings on both the dishwasher and the clothes washer.

In places where natural or bottled gas is competitively priced, most of the 73-kilowatt-hour-per-month load cited for the dryer in House 1 could be eliminated through the installation of a gas dryer. This might not be a move you would consider if you were staying with utility power and the gas and electric rates were close to equal. But the elimination of heavy loads such as a dryer from a photovoltaic system can have an extremely beneficial effect upon the sizing and cost-efficiency of that system.

We might also assume that in House 3 there is less reliance on appliances and gadgets, resulting in a 15 percent reduction in the base kilowatt-hour-per-month figure after the major-appliance deletions have been made. Improvements in siting, design, and insulation—as well as night-setback thermostat settings—combine to make possible a 20 percent reduction in the amount of electricity required by the ignition and pumping functions of the conventional (backup) heating system. Finally, we deduct 20 percent from the base kiilowatt-hour-per-month figure for lighting requirements. This savings is due to the use of passive design techniques that allow a greater use of natural light, as well as to the installation of fluorescents wherever practical and the conscientious switching off of lights when not in use.

Table 6-4 starts with the base monthly figures for House 1, then makes the various subtractions mentioned and presents the final totals for House 3, the energy-conserving house.

Table 6-4. Electrical Consumption in House 3

	Winter (kWh/mo)		Summer (kWh/mo)		Spring or Fall (kWh/mo)	
	490.2*		487.2*		401.2*	
	−54.0	dryer & dishwasher reduction	−86.0	air conditioner reduction	−54.0	dryer & dishwasher reduction
	−17.8	heating reduction	−54.0	dryer & dishwasher reduction	−15.0	lighting & appliance reduction
	−15.0	lighting reduction	−15.0	lighting reduction	N/A	
	−41.0	misc. small appliance reduction	−41.0	misc. small appliance reduction	−41.0	misc. small appliance reduction
Totals	362.4		291.2		291.2	

*kWh/month for House 1, including lighting.

For easy comparison, table 6-5 displays the figures for all three houses together.

It is obvious that extravagant use of electricity could result in a monthly kilowatt-hour sum well above the "high" estimate of House 2, especially if electric resistance heating were used; that frugality could yield savings much greater than those represented by the "low" figure; and that even the base calculation could vary significantly depending upon climate and the habits of the consumers. Thus, these figures cannot claim to represent some kind of "average" or "norm." What they *are* intended to show is that each individual load in a residential system contributes to the overall load to an extent determined by its current draw and duration of use and that the cumulative figures can vary tremendously depending upon the efficiency of appliances and the amount of time they are in use. This trans-

Table 6-5. Total Electrical Consumption in Houses 1, 2, and 3

	Winter (kWh/mo)	Summer (kWh/mo)	Spring or Fall (kWh/mo)
House 1 (base)	490.2	487.2	401.2
House 2 (high)	1,257.1	1,407.7	1,132.5
House 3 (low)	362.4	291.2	291.2

lates into hard numbers when the monthly bill arrives, and the numbers have dollar signs in front of them. Unfortunately, the contributing components—such as those shown in these tables—are lost in the total shown on your electric bill.

If you are considering the installation of a residential PV system, either as a retrofit to your present home or to power a new one, the monthly kilowatt-hour consumption recorded in your present electric bill takes on added significance. Reviewing this basic information will provide you with a good understanding of your present electricity requirements and serve as the starting point for determining the size, sophistication, and cost of your PV system.

For those who are working within the limits of some financial constraints, and that includes most of us these days, the problem at hand breaks down into two important questions: (1) how much electricity do you really need? and (2) how much electricity will the PV system that you can afford to install produce? Clearly, a balance must be struck between the two, and, as we shall see, the answers to the first question are generally more susceptible to compromise than the answers to the second.

Conservation Is the First Step

You can power anything you want with photovoltaics. However, when financial considerations become the dominant factor in the system design equation, it quickly becomes clear that it is easier and far more cost-effective to conserve electricity than it is to produce it. This does not necessarily mean doing without. What it does mean is doing more with less, by using intelligent design and careful equipment selection to deliver an equal or better living environment and lifestyle while consuming less energy.

First, you must decide which domestic functions are *most appropriately* powered by electricity. In many cases, the answers will be inevitable; there simply aren't any propane-fired blenders on the market. But elsewhere there are often choices. As noted earlier (table 6-1), a typical clothes

PHOTO 6-1
Photovoltaic modules such as this one by Pulstar Corporation, mounted on the left side of these solar thermal collectors, now power the pumps and controls in many solar heating systems.

dryer exerts a monthly demand of 73 kilowatt-hours. Aside from the option of solar-drying with a clothesline whenever possible, you can select a gas-powered dryer rather than an electric model if gas is available. Most gas dryers now have electronic ignition, which conserves gas and thus further reduces overall energy requirements.

The domestic hot water system offers another opportunity for load reduction. You will recall that in House 1 (moderate-use house), electric water heating was not considered as part of the load. House 2, by contrast, had an electrically heated, 3-kilowatt, 80-gallon water tank that was responsible for the consumption of 410 kilowatt-hours each month. (A quick-recovery tank might easily be rated at 4.5 kilowatts, although on the average it would draw current over a shorter duration and require about the same number of kilowatt-hours per month.) If we add two or more members to the family and suppose a 120-gallon tank, the consumption of electricity goes up even more.

A gas-fired water heater is the logical conventional alternative, provided gas is readily available. Or you might choose a heat pump that draws heat for DHW from a source such as the outdoor air or the air in your basement. However, the most attractive option by far is a solar DHW system, which requires only a very small amount of electricity to operate the circulating pumps that transport water (or a heat-transfer fluid) to and from the collector and storage tank. There are direct photovoltaic-powered solar DHW systems and there are also passive solar domestic water heaters available that require no electricity at all. Even though the amount of backup energy required in a well-designed solar DHW system will be insubstantial, the use of the conventional gas-fired (or oil-fired) water heater for backup is still preferable over electricity. If the house uses the solar system for its space heating as well as for DHW, it would make sense to use the same backup system to serve both the space and DHW functions.

Space Heating and Cooling

Electric resistance space heating, of course, draws such large amounts of electricity that it was not even presumed to be part of the electric load in the energy-conserving House 3. The conventional "all-electric house," in which an electric furnace or baseboard resistance units provide 100 percent of the space-heating requirements, has become economically undesirable in all but those few areas of the country blessed with an abundance of very inexpensive hydroelectric power and/or a very mild climate. Still, many people use portable electric space heaters for increasing comfort in those corners of their homes that are not adequately served by their central systems or as a means of allowing lower overall thermostat settings. This is not a wise use of utility-supplied electricity, and it is extremely hard to justify if the electricity consumed is generated by photovoltaics.

The preferable option, and by far the most practical and cost-effective one, is to design a well-insulated passive solar home that requires no conventional cooling and a minimum of supplemental heating. In the case of an existing home, it is far more economical to retrofit insulation and weather stripping and even replace old windows and doors to reduce heat-

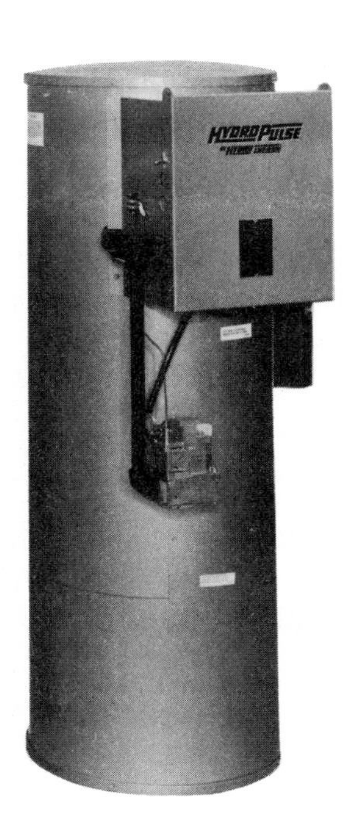

PHOTO 6-2

A family of high-efficiency, gas-fired pulse combustion boilers manufactured by Hydrotherm Corporation.

ing requirements and stop air infiltration than it is to continue paying high heating bills. This is true regardless of the type of heating system you have.

Whether you are planning a PV system as part of a new home or upgrading the energy use profile of your existing home in preparation for a PV system, the best options for auxiliary space heating are the new, high-efficiency gas- or oil-fired boilers and furnaces and/or an airtight, high-efficiency woodstove.

Heat exchanger**—a device for transferring heat from warm indoor air to cool outdoor air during the ventilation process**

Recent advances in gas and oil furnace and boiler technology have been truly remarkable. Increases in efficiency have largely centered around the use of heat exchangers, in which hot exhaust gases that would otherwise be vented directly to the outdoors are instead condensed and cooled before discharge. The heat thus recaptured (exhaust gases in a conventional furnace can reach 300° to 500°F) represents as much as 25 percent of the energy value of the fuel consumed—energy that would otherwise be lost up the chimney. These condensing furnaces, so called because they condense the water vapor in the exhaust gases, can approach 96 percent efficiency and save homeowners up to 35 percent of space-heating fuel costs. As of this writing, the leaders in high-efficiency residential furnaces are Lennox, Amana, Heil-Quaker, and Hydrotherm.

The airtight woodstove has by now become a well-known fixture in the colder regions of North America and needs little introduction. These stoves squeeze more heat energy from wood by controlling the amount of air available for combustion, thus preventing rapid, inefficient consumption of the fuel. They are called airtight not because they admit no air (if that were the case, there would be no combustion at all) but because they represent a radical departure from the loosely constructed, air-gulping stoves and open, damperless fireplaces of the past.

In extremely warm climates where even the best-designed passive solar houses require some mechanical cooling, an evaporative cooler might well be the answer. The blower of an evaporative cooler moves outside air through a wetted surface. As the water evaporates from this surface, heat is removed from the airstream and the air is cooled. The humidified, cooler air displaces warm indoor air, which is exhausted to the outdoors. Although this method of cooling is not very effective in areas with extremely high relative humidity, it is ideal in regions where the humidity rarely exceeds 50 to 60 percent because it is much more energy-efficient than compressor-driven refrigeration. State-of-the-art evaporative coolers can cool a well-insulated, 1,500-square-foot house with only 240 watts of power. Some units are designed to operate on direct current and be powered directly by the photovoltaic system, eliminating inverter losses. In comparison, a central air conditioning system can draw 2 to 4 kilowatts of power. (I will have more to say about evaporative coolers later in this chapter in the section on DC-powered appliances.)

Lighting, like indoor climate control, is another integral part of any home's energy profile, and it is one that bears close investigation if photovoltaics is to be considered. Because of the sizable amount of electricity that they convert into heat, incandescent bulbs are far less efficient and more expensive to operate than fluorescent lights. The catch is that incandescents have traditionally been far kinder to the eye and to the people and objects upon which their light falls. Fluorescents have a reputation for casting a cold, inhospitable light and for discouraging people from looking in mirrors. But there have been significant recent improvements in the quality of fluorescent light. The new, warm white fluorescent bulbs bring out the colors of the spectrum more naturally, while retaining their energy-efficient qualities. In parts of a house that require lighting for long periods of time, these bulbs should provide an acceptable alternative to incandescents, and their efficiency will be reflected in reduced kilowatt-hour demand.

To gain even more efficiency from your PV system, consider running your lighting—both incandescent and fluorescent—directly on your DC power source, reducing the use of an inverter and its associated internal conversion losses. Incandescent bulbs operate well with direct current, and DC ballasts are now available for virtually every fluorescent fixture. There are even many new fluorescent bulbs that will screw directly into a standard incandescent base.

The Importance of Appliance Efficiency

Ever since the era of electrical household conveniences dawned at the beginning of this century, consumers have been far more concerned with the quality, performance, and initial cost of appliances than the efficiency with which they used electricity. This attitude began to change during the energy crises of the 1970s, when engineers and laypeople alike discovered what would have been apparent all along had it not been for the complacency induced by low electric rates and inexpensive fossil fuel—that home appliances can be energy guzzlers, just like inefficient automobiles, and that there was much room for improvement.

Fortunately, energy-conscious consumers can take advantage of significant advances in appliance efficiency made during the past decade. Shopping for energy-efficient appliances is made easier by the efficiency ratings and estimated annual operating costs mandated by the Federal Energy Policy and Conservation Act of 1975. Energy Guide labels are required on furnaces, refrigerators, refrigerator/freezers, water heaters, clothes washers, dishwashers, and room air conditioners. According to the U. S. Department of Energy, these major appliances consume 73 percent of the energy used by the residential sector in the United States and can vary significantly among brands in terms of operating costs (kilowatt-hours consumed).

When you shop for a refrigerator, the Energy Guide tag on the door of each model will tell just how much an average year's use will con-

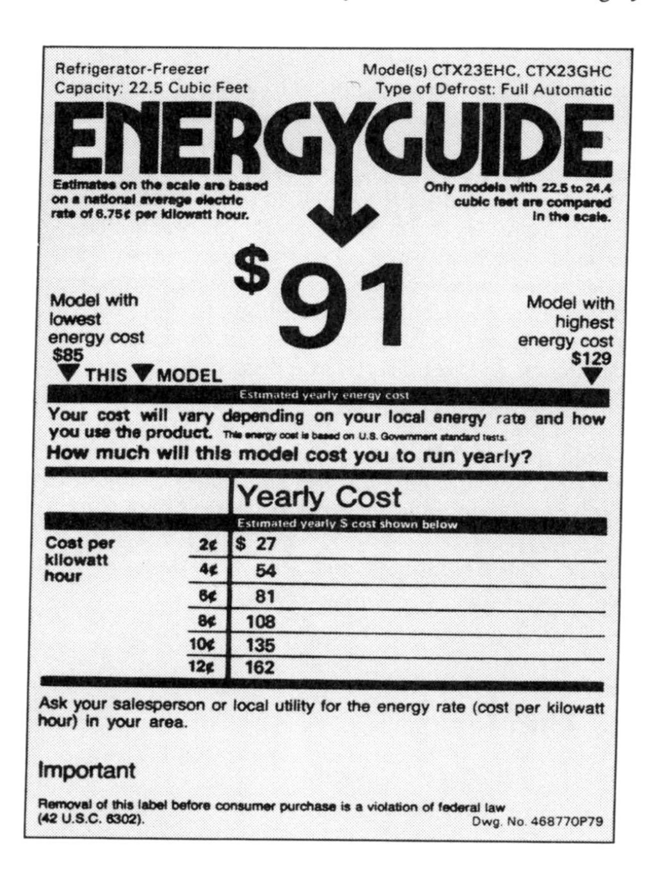

FIGURE 6-1

A sample Energy Guide label for a refrigerator, showing estimated yearly cost.

tribute to your electric bill. This information will be based upon a certain kilowatt-hour charge, which may or may not approximate what you are paying for utility power in your area. Simply divide the annual dollar figure by the cost per kilowatt-hour to arrive at the appliance's annual kilowatt-hour use. Compare that to the estimated kilowatt-hour use of other refrigerators to determine which is the most efficient model.

Room air conditioners are sold in a variety of sizes, each rated by the manufacturer according to its capacity as measured in Btu's (British thermal units) per hour. When buying an air conditioner, you should check the Energy Guide label for its Energy Efficiency Ratio (EER). The EER is simply the ratio of energy input to energy output—in this case, Btu displacement. The formula is EER equals Btu/hr divided by wattage. For example, an air conditioner that produces 5,400 Btu per hour at 600 watts has an EER of 9. The higher the EER number, the better (more efficient) the unit is.

Although a highly efficient air conditioner is likely to be more expensive than a less efficient one, the savings in electricity during the life of the appliance will far outweigh the initial higher cost. A good rule of thumb is to buy only an air conditioner with an EER of 8 or higher. In a 1983 report issued by the American Council for an Energy-Efficient Economy, the average EER among machines rated "most efficient" in the

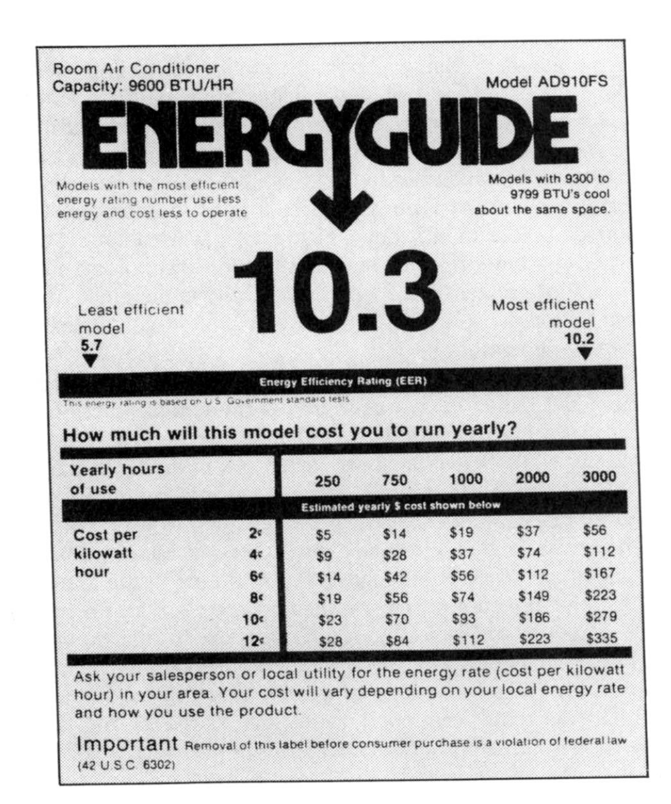

Room Air Conditioner
Capacity: 9600 BTU/HR
Model AD910FS

ENERGYGUIDE

Models with the most efficient energy rating number use less energy and cost less to operate

Models with 9300 to 9799 BTU's cool about the same space.

10.3

Least efficient model 5.7

Most efficient model 10.2

Energy Efficiency Rating (EER)

This energy rating is based on U.S. Government standard tests

How much will this model cost you to run yearly?

Yearly hours of use		250	750	1000	2000	3000
		Estimated yearly $ cost shown below				
Cost per kilowatt hour	2¢	$5	$14	$19	$37	$56
	4¢	$9	$28	$37	$74	$112
	6¢	$14	$42	$56	$112	$167
	8¢	$19	$56	$74	$149	$223
	10¢	$23	$70	$93	$186	$279
	12¢	$28	$84	$112	$223	$335

Ask your salesperson or local utility for the energy rate (cost per kilowatt hour) in your area. Your cost will vary depending on your local energy rate and how you use the product.

Important Removal of this label before consumer purchase is a violation of federal law (42 U.S.C. 6302)

FIGURE 6-2

A sample Energy Guide label for an air conditioner, showing estimated yearly cost and energy efficiency rating.

5,000-to-7,000-Btu category was 9. In the 9,000-to-11,000-Btu category, the average was 10.7. Airtemp, Fedders, White-Westinghouse, Emerson, and Friedrich are among the leading manufacturers of energy-efficient air conditioners.

Just how efficient can we expect common household appliances to become? The most efficient appliances today are certainly far more so than their counterparts of ten years ago; over the past two or three years, improvements have been less dramatic. Once innovations such as solid-state electronics, heavier insulation (in refrigerators and ovens), and moisture sensors in clothes dryers have been adopted, a point of diminishing returns begins to draw near. But, while the equipment that represents the cutting edge of appliance efficiency may now require quantum leaps of engineering or design to show significant additional improvement, there is still a healthy difference between the best units available and the often cheap-to-buy, expensive-to-run "guzzlers" that occupy the bottom end of the market.

Energy-Efficient Central Heating and Cooling

Energy efficiency is a concern that extends beyond the realm of kitchen, laundry room, and portable appliances to built-in heating and cooling equipment as well—even when electricity is not involved as a primary source of power. All gas- and oil-fired furnaces and boilers require some measure of electricity for ignition, and the newer, more efficient models on the market today cut down on even that small requirement. Electricity is a primary power source for heat pumps, a technology that has also made strides in efficiency. However, gas-fired central heating deserves our special attention, since new combustion technology makes possible an efficiency of greater than 95 percent, allowing the efficient use of a supply of fuel that can also be used to power the photovoltaic installation's backup engine-generator.

The big news in gas-fired heating equipment today is "pulse" combustion, as typified in products such as the Lennox Pulse furnace and the Hydrotherm Hydro-Pulse boiler. (The furnace is a device for heating forced air, while boilers heat the water in a hydronic space-heating system.) Pulse combustion involves the electronic ignition of a compressed mixture of gas and air. Exhaust gases resulting from the rapid burning of this fuel mixture travel instantly into a heat exchanger, where they heat either incoming air or boiler water, depending upon the application. Due to the negative pressure in the combustion chamber, a new supply of air and gas is drawn in and ignited, not by a new electrical arc from the spark plug but by residual heat from the initial combustion.

This "pulse" of repeated ignition and combustion occurs in a cycle of about 60 events per second, for as long as the thermostat requires operation of the unit. Another advantage of the furnaces and boilers that make use of this technique is the recycling of exhaust gases to heat the air or water medium on its return cycle. In addition to extracting more heat from the same amount of fuel, this procedure reduces the point-of-outlet temperature of the exhaust from a conventional 500° or 600°F to only 100° to 130°F, allowing fire-rated chimney flues to be replaced with ground-level outlet pipes, often made of PVC plastic.

The proof of the efficiency of these new furnaces and boilers lies in the percentage figures that represent annual fuel utilization efficiency, or AFUE. Conventional gas-fired boilers and furnaces had AFUE ratings of approximately 50 percent. Hydrotherm offers an AFUE of 90 to 91 percent in its Hydro-Pulse, while the Lennox Pulse furnace is rated at 96 percent. This 40 percent improvement in fuel efficiency more than compensates for the fact that these units cost approximately 30 percent more than conventional designs. Owners of the new combustion-efficient boilers or furnaces in colder parts of the United States can easily expect to recoup their additional investment during the first two winters, given the present natural gas price structure. Similar savings can be expected with propane.

AFUE—annual fuel utilization efficiency

Manufacturers of oil-burning heating equipment have also been working to improve efficiency. Energy Kinetics manufactures the System 2000 boiler, in which nearly complete combustion is achieved by the mixing of air and fuel forward of the injection nozzle in a high-temperature, ceramic-lined chamber. Exhaust gases immediately disperse into a spiral heat exchanger, transferring heat to the circulating water. As with most of these units, domestic hot water adaptation is available. The System 2000's AFUE is rated at 86 percent—again, some 30 percent higher than the old "guzzler" models. Also, not to be outdone by the gas appliance industry, the manufacturers of oil-fired furnaces have recently begun to introduce equipment with pulse combustion and condensing technology.

Heat Pumps

Heat pumps have been around for quite a few years, and their improved efficiency now makes them a good choice for inclusion in a medium-to-large-scale residential PV installation. It should be noted that a heat pump does not generate heat; it transfers it from one place (the *source*) to another (the *sink*), simply by playing upon the temperature differential between the source and a fluid cycled between condensation and evaporation. The heat pump extracts heat from its source of comparatively "cold" outdoor air or ground water by circulating a refrigerant gas through a condenser/evaporator cycle, much like a refrigerator or freezer.

In the *expansion phase* of its cycle, the gas in an air-source heat pump passes through an outdoor coil in which it cools to a temperature even lower than that of the environment from which it extracts heat. During the compression phase of the cycle, heat absorbed from the outside environment is given up indoors as the refrigerant passes through a freon-to-air coil at the core of the home's forced-hot-air heating system. In summer, the heat pump is used in reverse so that the indoor environment is the source and the air outside is the heat sink where unwanted heat is disposed of. The result: central heating and air conditioning, the two-for-the-price-of-one function that makes the heat pump especially attractive.

The heat pumps figuring most prominently in energy-efficient home design today are the *groundwater-source* or *earth-coupled* heat pumps, which are true geothermal energy devices. They extract heat not from steam-filled fissures deep underground but from groundwater at conventional well depth. This supply of water remains (below frostline) at a more or less constant temperature of 10°C (50°F) throughout the year, more than sufficient for successful heat pump operation in heating or cooling modes.

In a typical groundwater-source heat pump installation, the refrigerant, usually freon, is cycled between the ground level compressor and the bottom of a pipe sunk in the earth, ordinarily to a depth of several hundred feet. Firms offering heat pumps of this type include Hydron and Temp Master. Hydron advertises a coefficient of performance (COP) of 3.2 during the heating season, meaning that 3.2 units of heat are produced for each unit of electricity required to run the compressor on the pump. Results to date with groundwater-source heat pumps indicate that these devices may be by far the best, if not the only, justifiable method for harnessing photovoltaic power to the task of electric-powered space heating.

Coefficient of performance (COP)—**a measure of the amount of energy produced in relation to the amount of energy expended on energy production**

Direct-Current Appliances

Earlier I focused attention on the development of energy-efficient conventional AC-powered appliances. It is also important to note the special realm of appliances designed specifically for use with DC systems. Remember that, except for universal motors and incandescent lights, AC and DC equipment is not interchangeable. Though the scope and diversity of AC equipment is far more abundant, a great many electrically powered items are now available in DC versions.

In general, when we talk about DC appliances for use with modestly sized residential power systems, we are talking about 12-volt equipment. There are, of course, exceptions to this generalization, most of which fall in the 24- and 48-VDC range. A device called a DC-to-DC converter, which functions much like an inverter, handles the interface between differing system output and appliance voltage requirements. There are converters that will step up DC voltage (say, from 12 to 24 volts), and others that do the job in the other direction. As with DC-to-AC inverters, DC-to-DC converters within a system can be dedicated to the needs of one or more appliances, while the remainder of the loads operate at the level of system output.

DC-to-DC converter—**a device used to step DC voltage from a power source up or down as needed to match the voltage requirements of a piece of equipment**

Lighting

There are, of course, two basic types of artificial lighting for home use—incandescent and fluorescent. In an incandescent bulb, a tungsten filament glows within an inert gas (argon). In a fluorescent tube, the source of illumination is an electrically charged gas. Both fluorescent and incandescent lighting can be compatible with a DC system. There are, however, certain things to keep in mind in planning the adaptation.

An incandescent fixture can operate equally well on either alternating or direct current. The same is true of the bulbs they use, with one important difference. In a home electric system served by battery storage, there is no provision for output voltage regulation. In a 12-VDC system, fluctuation is minimal, but in 48- or 120-VDC setups it could be significant enough to overpower incandescent bulbs and seriously affect their longevity. For direct current use, then, bulbs with a higher voltage capacity (for example, 130 VDC for a 120-VDC nominal system) are recommended.

The same caveat applies to fluorescent bulbs used with DC systems. But here things get a little more complicated. Not only do the bulbs have to be compatible with the system voltage, but the fixture does as well. Incandescent bulbs are self-contained, but a fluorescent is ignited by means of the ballast in the fixture. A DC fixture with the correct voltage rating is thus required.

There are plenty of DC-compatible light fixtures available. REC Specialties, Inc., manufactures a line of fixtures called Thin-Lite, originally designed for wall or overhead mounting in recreational vehicles but suitable for similar installation in stationary locations. The units range from 8 to 32 watts, and all are 12 VDC. Replacement ballasts are available. The same firm sells a low-pressure sodium vapor light, 18 or 35 watts at 12 VDC, for outdoor use. Solar Electric of Santa Barbara supplies both 12- and 24-VDC fluorescent fixtures in wattages ranging from 17 to 36.

In addition to the stationary fixture, it is also possible to buy plug-in, high-efficiency lamps. Photron, Inc., distributes CAE's Littlite, a gooseneck, high-intensity lamp with a miniature quartz-halogen bulb. It runs on 12 VDC, operating at only 3.5 watts. Photron also carries a full line of DC receptacles, switches, and plugsets made expressly for residential power systems. The use of such equipment is especially important in dual-system

PHOTO 6-3

A wide variety of switches and receptacles specifically rated for direct current are available from suppliers such as Photron, Inc.

houses, where there is both AC and DC service. The possibility of matching up the wrong lights or appliances with the wrong receptacles is prevented by the DC plug/receptacle configuration, which makes the insertion of an AC device impossible.

Water Pumping and Delivery

Pumping water is one of the oldest applications to which PV-supplied electricity, particularly in the form of direct current, has been applied. Remote locations requiring water for irrigation and livestock were, and continue to be, ideal places to put photovoltaics to work. As this market developed, so did DC pump technology, so that today's residential DC photovoltaic system installer has a variety of pumps from which to choose. These include jet pumps, submersibles, jack and recirculating pumps, progressive-cavity designs, and water pressurization systems.

Among the manufacturers who advertise their pumps as being suitable for service with DC systems are the Dinh Company and A. Y. McDonald Manufacturing Company. The Dinh Company offers a low-cost, shallow-well pump advertised as capable of providing a minimum of 500 gallons per day, and McDonald markets a line of ½- through 1½-horsepower, self-priming centrifugal pumps.

Sun Amp Systems, Inc., manufacturers of a line of progressive-cavity water pump systems, advertises flow volumes of from 0.25 gallon per minute (GPM) to 100 GPM. Incremental increases in output are made possible with modular expansion of the PV array. Pump motors serving this system are available in a range of sizes from 1/20 to 20 horsepower.

Some manufacturers offer DC pumping systems that are specifically designed for use with swimming pools. One is the Sun Swim, marketed by Photocomm, Inc. Available in ½-, 1-, and 2-horsepower models, the Sun Swim features a controller that makes either AC or DC operation possible.

One DC pumping application that especially interests PV homeowners has to do with active solar domestic hot water systems. Since a PV home is often likely to also incorporate solar DHW, a DC pump can come in handy—either as part of an all-DC installation or as a means of taking one more load off a DC-to-AC inverter. Two promising solar-DHW DC-powered pumps are the Pulstarter, from Pulstar Corporation, which is designed to pump propylene or ethylene glycol in a closed-loop system, and the two 12-VDC units made by March Manufacturing, Inc. These pumps, rated respectively at 3.1 and 5.5 GPM, are for use in open-loop drainback or draindown systems.

The job of water purification can also be accomplished in an all-DC photovoltaic electric system. Some PV-powered options for water purification include multistage filtration systems such as the Seagull 4 line manufactured by General Ecology, which utilizes submicron filtration to remove waterborne pathogenic bacteria, *Giardia,* many organic chemicals, and even radioactive fallout. The system works on as little as 20 pounds of line pressure, which can easily be provided by a PV-powered pump.

Refrigeration

Solar refrigeration, which at first glance sounds like the ultimate contradiction in terms, was actually one of the earliest PV applications to be explored. The reason for this was the need for cold storage of medicines in remote parts of the world.

The conventional AC refrigerator is not a wise choice for use with SA photovoltaic electric systems, drawing as it does from 3 to 8 kilowatt-hours per day. For this reason, many PV homeowners opt for gas-powered refrigerators, although highly efficient DC models have been developed. DC-powered refrigerators are a good alternative to propane-powered models, either in all-DC situations or in homes where the aim is to put one less major load on the DC-to-AC inverter.

Norcold makes an 8-cubic-foot refrigerator (available from Solar Electric of Santa Barbara) that can be switched automatically to operate on either 12 VDC or 120 VAC. In either mode, the manufacturer claims that this appliance draws only 800 to 900 watt-hours per day. Another handy, small DC refrigerator is the Arctic Kold, a 5.3-cubic-foot, 12-VDC model available from Atlantic Solar Power, Inc.

PHOTO 6-4

This 16-cubic-foot, DC-powered refrigerator/freezer is manufactured by Sun Frost Corporation.

Larger DC-powered refrigerators are available from the Sun Frost Company. Their Model RF-12 has 12 cubic feet of storage area, and for those who require an even greater capacity, the Sun Frost Model RF-16 features 16 cubic feet of storage. Both models are available in either 12 or 24 VDC and are rated by the manufacturer as consuming only 13 and 15 kilowatt-hours per month, respectively. This represents about one-ninth of the average electricity used by a conventional AC refrigerator. The Sun Frost is a top-quality unit. It is rather expensive, but it is definitely the best way to go if you have decided to power your refrigeration needs with photovoltaics.

I have heard of several instances where Sun Frost units have been modified to receive supplemental cooling through the use of a heat pipe in cold climates during the winter. (A heat pipe is a device, usually filled with a refrigerant such as freon, that efficiently conducts heat from one point to another without the need for additional energy such as that used to drive a compressor.) Since winter is the time when the residential loads are highest and the photovoltaic array output is lowest, this approach is a perfect way to "have your cake and eat it, too" by installing a small PV system and still enjoying PV-powered refrigeration all year. I am told the manufacturer is helpful in providing information and support for such modifications to its equipment.

Cooling and Ventilation

Direct-current fans fall into two basic categories: vent fans, designed to exhaust warm air through attic spaces; and circulating fans, which not only cool in summer but help with winter heating by keeping warm air from accumulating near ceilings. Direct-current vent fans are available in a variety of sizes and are generally rated in terms of the cubic feet of air per minute (CFM) they are capable of moving. Major sources

CFM—**cubic feet of air per minute, a common measure of air movement**

for these fans include Solar Electric of Santa Barbara, Parker McCrory Corporation, and Solarex Corporation. Attractive DC interior circulating fans in the popular, wooden-bladed, Casablanca style are sold by Solar Electric and by the William Lamb Corporation.

The Nordika adiabatic evaporative cooler, available from Solar Electric of Santa Barbara, is built to operate on 48 VDC. Its fan circulates up to 2,200 CFM, which its manufacturer claims is sufficient to cool a well-insulated 1,500-square-foot home. The William Lamb Corporation offers two evaporative coolers operating on direct current; one is rated at 1,800 to 2,000 CFM and the other, larger model at 5,000 CFM. Similar units are available from Photocomm.

The highly efficient heat pumps manufactured by the Dinh Company especially for use with photovoltaics are a relatively new innovation. The two-speed, water-to-air devices feature a compressor and blower powered by a highly efficient DC motor. Two models are available: the WA-HP 1.0, which delivers 14,000 Btu per hour heating and 13,000 Btu per hour cooling, and the WA-HP 1.5, which delivers 19,000 Btu per hour heating and 18,000 Btu per hour cooling. The units deliver a coefficient of performance of 4.5 and are capable of operating either on 100 percent PV power or a combination of conventional AC power with a PV assist. The machines also feature domestic hot water generation through an optional desuperheater installed on the discharge side of the compressor, which works in both the heating and cooling mode.

Home Entertainment, Appliances, and Power Tools

The photovoltaic homeowner who has opted for a DC system does not have to do without modern home entertainment devices. Hitachi offers a 9-inch portable color TV capable of running on either 120 VAC or 12 VDC. On direct current, it consumes only 35 watts. The same company manufactures a combination tuner/speaker package with the same AC/DC capabilities. (A turntable would have to be operated on 120 VAC, unless a DC substitution of the conventional synchronous motor were made.) For camera enthusiasts, there is even a 12-VDC slide projector manufactured by Kindermann and available through Atlantic Solar Power, Inc. It can take either circular or universal magazines and uses a 12-volt, 20-watt halogen lamp.

The list of DC appliances grows each year. As of this writing, it includes washing machines (the ITT Mini Wash is a 12-VDC unit that holds up to 4½ pounds of clothes), vacuum cleaners and soldering irons, blenders and toasters, shavers and hair dryers, electric frying pans and coffeemakers. There is even a chain saw, the Minibrute, that can run off any 12-VDC system and easily clear away light brush and small trees. The washer and chain saw are available from Atlantic Solar Power; the other devices mentioned are cataloged by Alternative Energy Engineering of Redway, California.

Among the most useful devices on the DC equipment roster are simple electric motors of the type made by Honeywell in voltages from 12 to 90 VDC and power ratings of ⅛ to 1 horsepower. These can be used to replace AC motors in a wide variety of applications: pumps, washing machines, gas-fired clothes dryers, evaporative coolers, and stationary shop tools such as lathes, drill presses, band saws, and table saws.

It would be possible to continue talking at great length about the tools and appliances built to run on direct current electricity. But even without cataloging each device, the point should by now be clear: the DC lifestyle need not be one of forced austerity. Many of the conveniences associated with alternating current can be enjoyed with direct current if one makes the right equipment purchases and employs a little creativity. Meanwhile, the list of DC-powered options grows larger every day.

Load Management

There is more to a house's electrical load profile than the simple sum of all of its current-drawing appliances and equipment. Not all of the appliances are in use at the same time, nor do the many possible combinations result in a uniform pattern of electricity use. Indeed, a graph showing the rate at which electricity is consumed each day—both in the individual home and over entire power grids—is a hill-and-valley affair, as can be seen in figure 6-3. The high points on the graph are the *peak loads,* or times of greatest electrical demand.

The subject of peak power demands is of tremendous importance to homeowners operating independent power-generating systems, as indeed it is to the electric utilities. During peak load periods, the question is not one of how much power is generated each day but of *how much is available at a given time.* To an electric utility, this is crucial because a shortfall can result in spot outages and brownouts. To the owner of a stand-alone PV system, the price paid to meet such time-specific shortfalls is a larger and more expensive inverter and possibly a larger battery storage bank and/or a larger backup generator. For the line-tied installation, it will almost certainly involve increased dependence on utility power.

For most houses, the patterns of peak electrical demand are reasonably predictable. Statistics drawn from the experience of public utilities

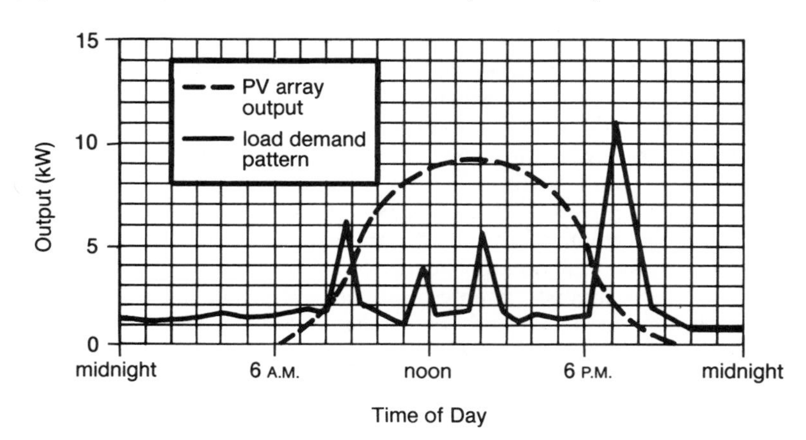

FIGURE 6-3
A graph of the daily output of a photovoltaic power system superimposed over a daily load profile for a "typical" residence.

show that the first of several daily peaks occurs between 7:00 and 9:00 A.M., when people get out of bed, turn on the lights, the radio, or the television, take a shower, and cook breakfast. A somewhat smaller peak corresponds to midday, although this phenomenon is declining along with the one-income family. There is a moderate midafternoon peak, largely attributable to clothes dryers and other appliances but enhanced significantly in summer by air conditioners struggling against the warmest part of the day. The heaviest period of peak demand takes place between 5:00 and 9:00 P.M., when people cook dinner, wash dishes, watch television, take baths, and perform the day's remaining chores that involve electric appliances.

Peak-hour loads, such as those briefly outlined above, are only one aspect of the complete load profile. At certain times during each peak hour, demand will be even higher and will be reflected in *peak-minute loads.* And within each peak minute, there are even *peak-second loads,* which are associated with the surge demand imposed by the starting of *induction motors* such as those in refrigerators, air conditioners, and stationary power tools.

In PV systems, the maximum output is achieved during bright sunshine at midday. The approach, attainment, and decline of this optimum level of solar electric generation is best represented on graph paper as an approximate bell curve. Superimposed over a chart showing the ups and downs of hourly demand, it engulfs certain peaks and only partly surmounts others, as can be seen in figure 6-3.

It would all be a lot simpler if the peaks and valleys of on-site electrical generation matched the peaks and valleys of demand. But this is almost never the case with a home power system. Assuming that the various components of the system have been correctly sized in the first place, the battery storage bank will act as a *flywheel* to equalize output with demand. Or, in a UI system, the load will draw grid power to balance against the PV electricity sold to the utility during high-output/low-demand periods.

There is no point in trying to bludgeon the problem into submission simply by designing a photovoltaic installation that will carry a house through the greatest foreseeable periods of peak demand. It's prohibitively expensive, and even if you have the money, you probably don't have the roof area to mount the array. You should instead incorporate measures designed to moderate those demands. This is where *load management* comes in. Along with the sensible selection and use of efficient home appliances, the practice of load management is a sound method of assuring that each kilowatt-hour used is worth what you paid for it.

At its simplest, load management consists of a series of conscious decisions concerning when to add certain appliances or power-consuming functions to the total domestic load. For instance, you might decide not to run the clothes dryer, vacuum cleaner, and oven simultaneously, but stagger their use. You may already be familiar with this practice if your electric utility has *time-of-use rate structure* (TOURS), which encourages the consumption of electricity during off-peak times. This process of decision-making is called *voluntary* or *participatory load management.*

TOURS**—time-of-use rate structure, devised by utilities as an incentive to persuade customers to relegate many electricity-consuming tasks to off-peak hours**

This approach, however, works better with time-specific loads—such as appliances turned on and off manually, as needed—than it does with non-time-specific loads—such as central-system climate control, well-water pumping, and refrigeration. These latter functions are provided by equipment that does not have to be in operation at any given moment, as long as it runs for a certain amount of time over a given period. Since it is not standard practice, even among the most energy-conscious individuals, to travel about the house throwing the switches of heat pumps and freezers in order to run the electric broiler, engineers have devised a means of *involuntary* or *nonparticipatory load management*. Involuntary load management is based upon microprocessors or small computers programmed to create a hierarchy of load priorities.

Involuntary load management—the use of small computers to switch loads on and off according to a programmed hierarchy of priorities

It was not photovoltaics but simple economics that created the impetus for development of load-management microprocessors. The prevalence in some areas of the TOURS rate structures makes the controlling of peak-demand patterns an extremely attractive objective, particularly for commercial users of electricity. Once the technology had been refined for large-scale installations, the practicality of computerized load-management modules for residences—particularly those equipped with time-sensitive alternative energy systems such as photovoltaics—was quickly tested and proved.

How does a load-management microprocessor work? Typically, it is a programmable device that monitors overall electrical demand. It constantly compares the extent of this demand with a demand limit preset by the user. When the microprocessor senses that actual demand is about to exceed that limit, it refers to its programmed hierarchy of non-time-specific loads and turns them off as necessary. The loads remain off for as long as is required to prevent excess peak demand, within a programmed window of minimum and maximum times.

One home load-management microprocessor, the Savergy SC-112, can be programmed to control loads according to any of three user-specified strategies. The first of these is termed by the manufacturer the "fixed-priority strategy," and consists of a simple eight-load hierarchy. The second is the "rotate" strategy, in which all monitored loads are turned off in a sequential cycle every 30 seconds as necessary to keep peak demand below the set limit. In this mode, the first load off is the first back on-line at the conclusion of each cycle, unless circumstances require that all loads remain off for a brief simultaneous period. The third strategy combines the first two, assigning a hierarchy to three circuits while the remainder rotate in a lower order of priority.

The manufacturer of this microprocessor also offers a unit compatible with Commodore home computers, with software designed for either the basic load-prioritizing, peak-demand management function or for simple timing of separate loads.

The use of load-management microprocessors to monitor and regulate residential electrical demand is a rather new concept but one that makes excellent sense when the object is to keep minute and second peaks within reason. In conjunction with an inverter capable of responding adequately to such surge demands as are unavoidable, it can make possible the implementation of a smaller, more efficient, less expensive PV system.

Where the photovoltaic installation is small and custom-designed load-management devices constitute overkill, a simple relay has been known to do the task quite well. For example, I know of several instances where a relay was installed to come on whenever a home's water pump starts up. This relay drops some large AC loads that would compete for power off-line and allows the inverter's full output to be available to start the water pump. When the pump has filled the pressure tank, it shuts off and the relay drops back to normal and reconnects the other loads. The precise matching of the resource to the job at hand is perfectly in keeping with the spirit of the decision to use photovoltaics and the innate elegance of the technology itself.

In chapters 1 through 6 we have become acquainted with the photovoltaic process, the hardware that makes it possible, the job it has to do, and the ways in which we can make that job easier. The remainder of the book will be devoted to a discussion of system topology, design, installation, and maintenance.

CHAPTER SEVEN
UTILITY-INTERACTIVE PHOTOVOLTAIC SYSTEMS

Throughout the developed world, residents of all but the most isolated areas are able to take advantage of an electrical utility network that combines highly centralized generation with an extensive grid of overhead and underground transmission lines. Virtually without exception, the means employed by public and private utility firms to generate the power distributed through these lines fall into three categories: fossil fuels (coal, distilled petroleum, and, to a lesser extent, natural gas); nuclear fission; and the energy stored in moving water, known by the generic term *hydropower*. Fossil and nuclear fuels create the motion necessary to turn generating turbines by heating water until it becomes steam. With hydropower, the movement is implicit and need only be harnessed by turbines at the dam.

So far, the system has worked remarkably well. But the commitment to centralization implicit in it has, of late, given rise to a number of thoughtful criticisms. The first, expressed succinctly by energy analyst Amory Lovins in his book *Brittle Power* (Brick House Publishing Co., 1982), has to do with the inherent vulnerability of any complex and rigidly centralized utility infrastructure. Lovins and his allies argue that catastrophic interruptions of service such as those that occurred in New York City and throughout the northeastern United States in 1965 and 1977 are not flukes but built-in inevitabilities, and that acts of sabotage could easily wreak far more havoc than might result from such mere accidents.

The second argument voiced by those uncomfortable with centralized electrical generation concerns the accompanying forced dependence upon the fuels required to keep all but hydro plants in operation. Each fuel has its own special problems. The mining of coal is dangerous and/or environmentally disruptive and carries with it the nagging problem of acid rain unless expensive pollution controls are applied. Oil is subject to politically motivated supply interruptions, as we are all too aware of since that fateful autumn of 1973. As for nuclear power—the fact that the nuclear debate continues with such unabated vehemence is sure to preclude the widespread conversion of uranium and plutonium into kilowatt-hours.

One further problem with the centralization of generating facilities is that of capital costs. It takes a tremendous amount of money to put a new utility plant—nuclear or conventional—on-line, so the decision and the complicated financing arrangements attendant on it have to be made many years, often a decade or more, in advance of projected completion. To further compound this problem, these decisions are made on the basis of demand estimates that may or may not turn out to be accurate. Increasingly, they do not.

The solution proposed by a number of progressive energy strategists is the joint production of power by centralized generating facilities *and* a broad range of dispersed small-scale power producers distributed along the grid. Some optional sources of power being suggested include reactivated "low-head" hydroelectric sites that were earlier abandoned, wind turbines and wind farms, reclamation centers that turn urban refuse into energy, commercial and industrial cogeneration, electric generators fueled with biomass, geothermal power, and, of course, utility-interactive (UI) photovoltaics.

Of the renewable options, photovoltaics offers the most potential by far. Water power and (to a slightly lesser extent) wind power are highly site dependent. There are only so many sites at which conditions are suitable for the development of these resources. By contrast, photovoltaics has proven itself in applications from the equator to the Arctic. As the cost of PV electricity continues to come down, more and more UI applications will become economical and the world-wide electrical production from PV systems will grow geometrically.

One of the great attractions of photovoltaics is the siting flexibility that it offers, which allows generating facilities to be located close to the point of power use, thus minimizing losses associated with long-distance transmission. As long as an unimpeded southern orientation of the array is possible, special site requirements and/or preparation can be kept to a minimum.

Even in an urban environment there is plenty of space available that is suitable for the development of large-scale photovoltaic (PV) array fields. Modules could be mounted upon simple, lightweight space-frame structures built over parking lots at shopping centers, industrial parks, college campuses, and hospitals, providing shade while producing energy. PV systems could also be built in the airspace above superhighways and expressways, not to mention the thousands of square miles of underutilized flat roofs of commercial, institutional, and industrial facilities.

In this chapter, we will review the basic topography of a UI solar electric system, with special emphasis on the hardware needed to make the utility connection and on the details of the special relationship that exists between dispersed power producers and the utility companies. First, let's take a look at how the utility companies might be affected by this relationship.

Utility-Interactive Photovoltaic System Configuration and Components

From the outside, a UI photovoltaic installation looks much the same as a stand-alone (SA) system. However, there are fundamental differences in the internal relationship of the components, particularly with regard to the function of the inverter. In short, a UI system is not just an SA system without batteries.

As figure 7-1 shows, direct current (DC) electricity generated within the modules of the array flows through lightning protection and source-combining equipment directly to the inverter, which, along with its filters and controls, is referred to as the power conditioner in a UI system. At the power conditioner, the raw DC output of the array is inverted to alternating current (AC) electricity of the quality required for load consumption and grid interface as described in the section on UI inverters in chapter 5. At this point, depending upon the level of output and the fluctuating demand of the house loads, the AC power from the power conditioner flows either to the loads or—during times of excess output—to both the loads and the utility grid. If there are no on-site loads, all of the PV-generated electricity will flow into the grid.

So far, we have been following the progress of electricity in one direction—that is, from the array to the house loads and the utility grid by way of the power conditioner. But all the while electricity must also be able to flow *into* the house from the utility grid itself. This connection is necessary in order for the UI inverter to operate, as you will recall from the discussion in chapter 5.

With a UI photovoltaic system, there are then two paths by which electricity can enter the household circuitry leading to the loads: the out-

FIGURE 7-1
A block diagram of a utility-interactive residential PV system.

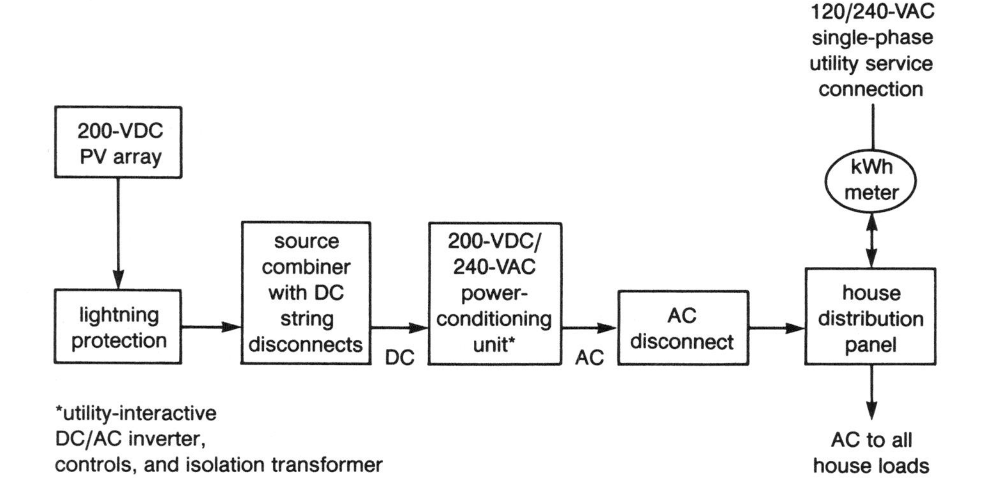

put line from the power conditioner and the line from the utility service entrance. Both can function simultaneously. Also, the utility mode can function separately without any contribution from the PV system, because utility power does not flow through the power conditioner to enter the house distribution. However, the PV side of the system cannot operate with complete independence from the utility. With present equipment, the utility requires that the PV system immediately disconnect from the grid whenever the grid fails. This eliminates any possibility of PV-generated power being fed out onto a downed line and endangering service personnel.

As we noted in chapter 5, there are two possible exceptions to this technical rule. You can install a second inverter capable of stand-alone operation, or you can install one of the new, semiexperimental, bimodal, self-commutated inverters capable of either line-tied or SA operation. However, these new inverters will not be accepted by the utilities until it can be shown to the satisfaction of utility engineers and the various safety and code organizations that these units will automatically disconnect from the line when there is a fault on the utility grid and will not send current out onto the lines when the grid is down.

Since there can be no photovoltaic contribution at night, and since an inverter kept in operation by incoming grid current but doing no work can represent a parasitic load in and of itself, most UI systems incorporate an automatic "night mode" sensing circuit, which deactivates the inverter during periods when PV generated-power is inadequate.

Metering, as we noted briefly in chapter 2, presents us with two basic configurations. If the utility that services your home approves, a conventional meter able to record power flows in both directions can be used. Otherwise, two ratcheted, one-way meters must be used—one to record power sold and one to record power bought. This is obviously necessary when there is a difference between purchase and sell-back rates. The metering options available to you will most likely be decided by your state Public Utilities Commission and the utility that serves you, based on the agreed-upon rate structure.

That completes our review of the basic configuration of a UI photovoltaic system. Inasmuch as its performance is no better or worse than the sum of its parts, the careful selection of the two most important components—the PV array and the DC-to-AC inverter—is crucial. We have covered the function and selection of UI inverters in great detail in chapter 5, so let's take a close look now at the process of sizing the array in a UI photovoltaic system.

Sizing the Utility-Interactive Array

There are three approaches to the sizing of a UI photovoltaic array. The first is simply to buy as many PV modules as you can afford, or perhaps have roof space to accommodate, and let them contribute what they will to the total load demand. The second is to let an experienced PV system engineer or designer do the calculations, based upon load-demand data that you provide. The third way is to do the figuring yourself.

There are several things lacking in the first approach. If the number of modules that can be purchased falls far short of what is required for a meaningful photovoltaic contribution, the array is likely to become little more than an appendage to what in effect will be a utility-supplied system. Conversely, as module prices fall, the "all you can afford" sizing method might easily result in module overkill—an excess generating capacity resulting in the sale of a significant amount of power to the utility, which will be hard to justify economically unless you have an especially attractive buy-back arrangement. So the advice is, don't just walk in to your local PV dealer and write him a check unless you already have a better-than-ballpark idea of how large an array is called for.

The second approach to array sizing is self-explanatory, and it is likely to be the safest and most thorough way to go *if* all the data is in order. Remember that microclimate figures into the picture along with established regional climatic information and that several visits to the site, in different seasons, are recommended if you don't have access to local climatic data.

This leaves us with the third, or do-it-yourself, method, the rules of which are handy educational tools even if the actual calculations in planning are to be left to a professional. The sizing procedure for UI systems offered here is based upon the material presented in chapter 3. As insolation input data, it uses average daily hours of peak sun expressed in kilowatt-hours per square meter (kWh/m^2) and taken at the tilt angle of the array.

Before we begin, we should note that there is an important difference between the requirements imposed by utility-interactive versus stand-alone photovoltaic system design. It comes down to a matter of leeway. With the SA system, the object is to match both array output and battery storage to load demand in such a way that basic loads are covered and the batteries are capable of handling peak demands as well as carrying the residence through nights and the number of bad-weather days figured into the storage capacity design. With the UI system, the designer is not bound so rigidly to satisfy the entire load with the PV array, because both storage and backup power are supplied by the utility grid.

Daytime solar fraction—the ratio between average daytime power consumption and average daily photovoltaic system output

Now don't get the wrong idea. The fact of the utility grid being ever at the ready to pick up slack does not mean that careful load management and array sizing—or the entire concept of energy conservation that was emphasized in chapter 6—should be ignored. If that is going to be the case, there is little sense in making the investment in photovoltaics in the first place. What it does mean is that, in the absence of batteries, a different concept of matching output to loads comes into play. It is expressed in terms of the *daytime solar fraction,* which is a simple ratio of the average daytime power consumption to the average daily PV system output, both expressed in kilowatt-hours.

Having thus defined the solar fraction for PV systems, the first thing we must do is unravel it a bit by reminding ourselves that averages exist only in the abstract and that there will be daily and seasonal fluctuations in available insolation and array performance and also in loads. A representation of an ideal average day would have system output approximating load demand for a *load/solar ratio* of 1. But there will likely be many days when PV output exceeds the home's coincident electrical demand, allowing sell-back to the utility, and also many times when electricity must be drawn from the utility grid to make up for shortfalls in PV system production.

Actually, things get even a bit more complicated. Our ideal load/solar ratio of 1 is deceptive in that it does not describe a simple no buy–no sell situation. If output is equivalent to load, the actual amount of load to be displaced by solar output in an average day is about 70 percent, leaving 30 percent "surplus" for sell-back to the utility. This is according to sample calculations done by Y. P. Gupta, S. K. Young, and others in their *Design Handbook for Photovoltaic Power Systems* (Sandia National Laboratories, 1981).

How do we explain these apparently contradictory figures? They result from the fact that system output and loads do not occur at the same time. Even if the raw kilowatt-hour figures on both sides match up, the times of peak solar intensity and peak loads do not. The 70 percent/30 percent split referred to is based on an analysis of typical residential load profiles, which show that enough peaks are "lopped off" by solar shortfall to bring net photovoltaic contribution down to 70 percent, while the opposite phenomenon—solar overcapacity in relation to loads—prevails long enough each day to allow 30 percent of site-produced power to be sent to the utility.

In sizing a UI photovoltaic system, the goal is to strike a balance between the undesirable extremes of too little and too much solar displacement of loads. It is a mistake to size the installation so that the solar fraction will be so low as to not effectively utilize the capacity of the smallest UI power-conditioning unit (PCU) available (which is 2 kilowatts), but it is equally misguided to try to reach too high in covering daily peaks. The continuum of array-sizing decisions for utility-interactive PV systems could be represented by a curve beginning at the point where output is so low as to be not cost-effective, reaching an apogee at the point of optimum return on investment, and trailing downward toward the point at which over-investment in generating capacity outweighs projected sell-back benefits.

It should be easy to see, however, why it would be futile at this juncture to assign specific numbers to the coordinates along the curve. There are simply too many variables for a general statement to be made. These include the costs of modules and other components, both of which are on a consistent downward slide; the purchase and sell-back rates for electricity, which vary from place to place and from year to year; and the sell-back rate structure itself, which in most states is designed so that the utility may pay less for each kilowatt-hour it buys than it charges for each one that it sells.

Remember that PURPA mandated the basing of sell-back rates on the "avoided cost" of generation, which can be interpreted to mean the "wholesale" cost of producing electricity (at the margin) as opposed to the "retail" cost of selling it. Our curve is at the mercy of all these variables and might easily continue to rise if photovoltaic costs go down and utility rates, both purchase and sell-back, go up.

In any event, there will always be a point that represents what Gupta and Young term "the most economical mix of load displacement and sell-back," beyond which "costs will increase at a faster rate than income and net benefits will decrease with each increase in capacity." With the many parameters dictated by component costs, utility rates, available array space, and local insolation, the homeowner may wish to work with a professional photovoltaic system designer to determine that optimum "economical mix" and build the system accordingly.

The array-sizing process begins with a tally of all load requirements in watt-hours or kilowatt-hours. If you are retrofitting a UI photovoltaic system to your existing house, the best source of your electrical requirements will be your present electric bill. However, whether you use your electric bill or establish the electrical budget for your new PV-powered home from scratch, remember that the amount of power lost in the operation of the inverter to deliver AC power from the PV system must be considered in your total load demand calculations. As a rule of thumb, you should subtract 10 percent (possibly a lot more, perhaps a little less, depending on the design and efficiency of the inverter you have chosen) of the estimated DC output from your array to account for internal inverter losses.

To start the array-sizing process, first calculate your estimated load requirements as outlined in chapter 6, then review the material on solar resource evaluation and array sizing in chapter 3. To obtain the necessary input data on solar insolation for your site, refer to tables of insolation data (such as those found in several of the books referenced in the bibliography) and find the average daily number of peak sun hours expressed in kilowatt-hours per square meter.

Utility-interactive photovoltaic homes rely on the grid for backup, but the use of the average peak sun figures should be considered if dependence on the utility is to be minimized. In most cases, calculations show that a satisfactory load/solar ratio can be achieved using the average yearly figure. However, if winter solar contribution is to be maximized, the array should be sized for the worst-case month and/or the altitude angle of the array mounting should be increased to latitude plus 10 or 15 degrees.

To determine the desired peak-wattage output for the array, divide your estimate of average daily load demand by the number of average daily peak sun hours given for your area.

Here is an example based upon the load demand estimate that we calculated for the hypothetical low-use house in chapter 6. Average monthly use (winter) is 362.4 kilowatt-hours. Therefore, average daily use ($362.4 \div 30$) is approximately 12.1 kilowatt-hours, or 12,100 watt-hours per day. (For simplicity, it is assumed that inverter losses are already incorporated in this figure.) Let's assume our house is located in the North Atlantic states or the upper Midwest. Consulting a source of insolation data, we find that the average number of winter peak sun hours per day is 3, giving us the equation:

$$\frac{12{,}100 \text{ Wh}}{3 \text{ hr}} = 4{,}033 \text{ W}_p$$

The array peak-wattage requirement thus derived is 4.033 peak kilowatts. To this number we must add some qualifications. We know that the power output of a photovoltaic module is dependent on sunlight intensity and operating temperature. We also know that the module performance data ratings provided to us by the PV manufacturers are based on a presumed average sunlight intensity of 1,000 watts per square meter and a module operating temperature of 25°C (77°F).

The average sunlight intensity on the photovoltaic modules on your roof will almost always be lower than this 1,000 W/m^2 due to atmospheric conditions and an angle of incidence less than perpendicular. Likewise, the average operational temperature for the PV array in the field will be higher than 25°C. We learned in chapter 3 that when peak-sun-hour data is used for array sizing, the correction for insolation intensity has already been made mathematically, since the data represents the "equivalent hours of full sun." However, correction for temperature is still needed when array sizing is done with the watt-hour method.

Therefore, before we divide our array peak-wattage requirement by the manufacturer's rated output of the module being considered to determine the number of modules needed to make up an adequately sized array, we must adjust the module's rated output to account for the actual operating temperature the array will experience. On our sample homesite in the North Atlantic states, the nominal operating cell temperature (NOCT) of our array is likely to be about 40°C (104°F) if we are clever with the design of our array mounting, and closer to 50°C (122°F)if we are not.

Since the peak output power (P_m) of a photovoltaic module drops by approximately 0.4 percent with each degree Celsius of temperature rise, we quickly see that module cooling is very important to consider when designing a PV system. The difference between a 40°C and a 50°C cell operating temperature is 4 percent of the array's total power output. (For ratings of specific manufacturers' products, see table 3-1.)

If we choose an ARCO Solar M-51 module with a rated peak output of 40 watts and de-rate it in relation to a presumed NOCT of 45°C (113°F)—20 degrees Celsius higher than the rated operating temperature—we must reduce the power output by 8 percent to account for the higher operating temperature ($0.4 \times 20 = 8$). Thus, we can expect 36.8 peak watts ($40 \text{ W}_p \times 0.92$) of *average actual power* per module available from the array on our house when we calculate array output using the watt-hour method. So then, our final array-sizing equation becomes:

$$\frac{4{,}033 \text{ W}_p}{36.8 \text{ W}_p/\text{module}} = 109.6 \text{ modules}$$

Our calculation shows that it will take 110 modules, at 40 watts each (based on a delivered power of 36.8 W_p each), to make up an array capable of satisfying the full load demand projected for our model house.

As we learned in chapter 3, to create a working array, photovoltaic modules must be electrically connected in series to form a string of modules, what the National Electrical Code (NEC) calls a *source circuit,* to obtain the desired DC input voltage for the inverter. Each of these electrically identical series strings of modules is then wired together in parallel to deliver the rated current of the array at the desired voltage. In practice, the actual number of modules specified for this array might be different from our calculation, because 110 may not be evenly divisible into the correct number of "strings."

To complete our sizing example, then, we need to determine the exact number of modules required in series to form a source circuit (series string) and next decide how many of these source circuits will be required in parallel to satisfy our load requirements. To do this, we must first determine the operating voltage of the photovoltaic array as dictated by the DC input voltage requirements of the power-conditioning equipment. So we must now choose an inverter.

If, as an example, we choose a Sunsine UI-4000 inverter (manufactured by American Power Conversion Corporation) for this system, we find from the manufacturer's specifications that this unit has a DC operating voltage window of 200 to 300 VDC, a minimum start-up voltage of 231 VDC, a maximum start-up voltage of 338 VDC, and nominal or preferred operating voltage and current of 230 VDC at 19 amps, or a maximum DC power input capacity of approximately 4,300 watts.

We have frequently noted the decrease in the power output of a module or array when temperature is higher and insolation less intense than standard test conditions. But keep in mind that module operating parameters can go up as well as down. For example, consider the hypothetical wintertime condition of cold outdoor temperatures coinciding with a fresh snowfall and broken clouds in the sky. The low ambient temperature reduces the operating cell temperature, which increases the operating voltage. The snow cover provides very good ground reflectance, enhancing the available solar insolation.

The broken clouds can further enhance the available insolation by creating a magnifying effect when the sun comes through a hole in the clouds, delivering the full direct component *and* a substantial amount of reflected diffuse radiation. This enhancement has been measured many times in the field at values of 1.5 suns and greater for short periods. The power-conditioning equipment specified must be able to handle the full extent of the array's operating range without overstressing the power-output electronics. Some inverters have internal controls that automatically limit or disconnect the DC array power input when the array operating voltage approaches the inverter's safe operating limit.

We learn from the manufacturer's literature that the ARCO Solar M-51 photovoltaic module has a rated peak power voltage (V_{no}) of 17.3 volts under test conditions of 1,000 W/m^2 solar insolation and a cell operating temperature of 25°C. We also confirm that this voltage varies by about 0.5 percent per degree Celsius.

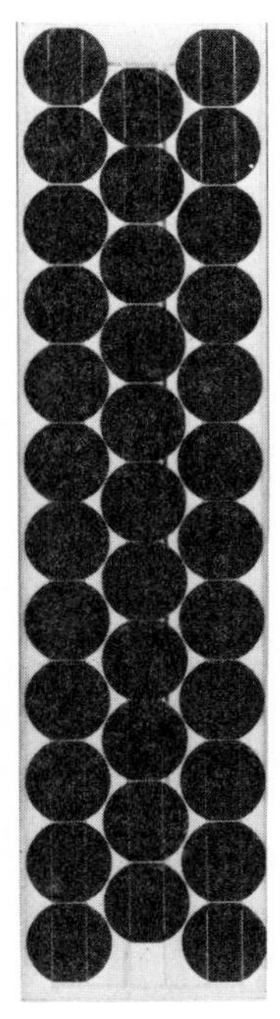

PHOTO 7-1
The M-51 photovoltaic module manufactured by ARCO Solar.

We stated earlier that we expect this module to have an average NOCT of 45°C under the conditions prevailing at our hypothetical homesite. Even though the operating conditions for the PV array on the house will change from season to season, and indeed from day to day, this NOCT more nearly represents the average conditions under which our array will be working than the manufacturer's data taken at 25°C, and we will use it for our calculations.

We form the following equation to convert the module's operating parameters from a cell operating temperature of 25°C to the more realistic NOCT of 45°C:

$$1 - [(45° - 25°C) \times .005\ V/°C] \times 17.3\ V = 15.57\ V$$

Thus, at a NOCT of 45°C, we have a decrease from the rated operating voltage at peak power to 15.6 volts.

On the other hand, under the previously mentioned wintertime conditions, the module would produce a *higher* peak power operating voltage. With an outdoor ambient temperature of 0°C (32°F) and an average NOCT of 25 degrees Celsius above ambient, we find that the module's nominal operating voltage would be the same 17.3 volts calculated on the basis of standard test conditions. If the insolation enhancement from snow and cloud reflection combine to deliver 1.5 suns (1,500 W/m^2) on the module coincident with this cold outdoor temperature, the peak power will increase proportionately with the sun intensity to 60 watts.

The difference in the V_{no} of a little less than 2 volts over the operating range of the module may not seem like very much change. However, it adds up rather quickly when you have series strings of 12 or 16 modules, and when combined with the changes in insolation levels can produce an array output to the inverter that is much higher than the average level for the system design.

Our chosen inverter has a nominal DC input operating voltage requirement of 230 volts. If we divide this input voltage by the V_{no} of an individual module at our estimated average conditions, here is what we find: 230 ÷ 15.6 = 14.74. This means that 14.7 ARCO M-51 photovoltaic modules will be required in series to make one string or source circuit with an output (V_{no}) of 230 VDC. Of course, we cannot use a fraction of a module, so we have to round off this number and decide that there will be 15 modules per series string for our array.

As we learned in chapter 3, the series string is the smallest increment of sizing possible for a photovoltaic array since it represents the minimum number of modules that we can assemble to deliver the required DC input voltage to the power-conditioning equipment. To maintain a consistent array operating voltage, all of the series strings that make up the array must work at the same voltage. Each string must therefore include the same number and type of modules as the others, and all of the modules in a string must be alike in their output.

Returning to our sizing calculation, we find that the figure of 110 is not evenly divisible by 15. This leaves us with two alternatives. We can use either seven strings of 15 modules for a total of 105 modules or eight strings for a total of 120. An even better approach might be to review our array calculations and consider using only 14 modules per series string and sizing the array at eight strings. Any one of these three options will perform satisfactorily with the DC-input voltage requirements of the power-conditioning equipment we have chosen. The final choice depends on what our goals are for the system and what our roof area and resources will allow.

The conservative choice would, of course, be the 120-module array, since electrical requirements in most households are more likely to increase than diminish as time goes on. This is especially true if maximum conservation measures are implemented initially in the original design.

There is another option. It is possible to initially install a smaller array of perhaps six or seven strings and—assuming plans for the space have been made—add more strings as time and resources allow. This scenario will involve some sacrifice of system efficiency, since the inverter will not initially be working at full rated capacity. However, the inverter we have chosen for this example is highly efficient throughout the entire power range and would lose only about 1 percent in conversion efficiency if it were operated with an array of six strings of 15 modules. Such an array would have a peak output of approximately 3,312 peak watts at a NOCT of 45°C.

The above estimates are based upon load projections for a residence that, though its energy use is conservative in conventional terms, nevertheless represents year-round occupancy and a fair level of use of electrical amenities. How do the figures look for a smaller house with less electrical demand? We'll work with a total monthly load estimate of 150 kilowatt-hours (5 kWh/day). Using the same figure of 3 for winter average peak sun hours, we arrive at the array requirement of 1,667 peak watts (5,000 Wh/day ÷ 3). If we choose the same 40-watt modules (de-rated to 36.8 watts), we find that 46 of them will make up an array adequate for our needs.

However, as we learned from the previous sizing example, we must size our array in increments of one string, and if we are to use the same 230-VDC array output voltage, that string length will again be 15 modules. Fortunately, our 46-module array is nearly evenly divisible into three strings of 15 modules each. If our house is to be used only during the warmer months, we can use the annual average peak sun figure of 4.5 hours, thus reducing the array peak-wattage requirement to 1,111 and the number of 40-watt modules required to 30, or two strings of 15 modules.

These examples illustrate the procedures used for sizing a UI photovoltaic array using the watt-hour method. The calculation methods presented will not deliver the accuracy of professional, computer-assisted sizing procedures; however, they will yield a better-than-approximate projection of array size and output and should be quite adequate for sizing the average line-tied residential PV system.

Provisions for Future On-Site Storage

In chapter 2, we presented the idea of a stand-alone/utility-interactive "hybrid" photovoltaic installation, incorporating both battery storage and a tie-in to the grid. To most people, this may seem difficult to justify at the present cost and level of development of PV technology. However, there are some applications where this type of hybrid system may make sense.

There are three possible reasons for including battery storage in what is essentially a UI system. The first is to add DC input capacity during peaks and/or evening hours, thus lessening the need for dependence on power from the grid. This is an active approach in which the batteries are cycled to some degree on a daily basis. The second reason would be to provide a complete, multiday backup power source ready to take over in the event of grid failure or other contingencies, such as a sharp rise in electric rates or the simple desire to stop paying monthly electric bills. This scheme necessarily involves a much larger battery bank, which would in effect be "floated" at or near full charge during periods of ongoing utility interaction, although it could also be used to a limited degree to address peak and evening loads. The third reason would be to completely eliminate all sell-back of photovoltaic-produced electricity to the utility. In this case the grid would be used only as a source of backup power during times when demand outpaced production.

With the first approach, a single conventional line-tied inverter can be used, since DC input is DC input whether it comes from the photovoltaic array or from the batteries, and the system is not necessarily being asked to function independently of the utility. In this case it is not truly a stand-alone/line-tied hybrid, except for the fact that greater leeway is provided on the site-supply side.

The power-conditioning options for the second, or full-storage, approach are a little more complicated. They include the following: (1) using two inverters, one stand-alone and one utility-interactive to match the system's alternative modes; or (2) installing a bimodal, self-commutated inverter capable of delivering an acceptable sine wave to the house loads even when it is isolated from the utility line. Again, it is imperative that such a device also incorporate a foolproof provision for complete utility cutoff lest power from the batteries or array find its way into the grid during a utility outage.

Without sell-back, the third system option would no longer fit the literal definition of "utility-interactive." Instead, the system configuration would take the form of the "utility-backed" system presented earlier in the book in figure 2-6.

In their *Design Handbook*, Gupta and Young point out that the inclusion of battery storage in the UI system topography also influences the matter of array peak-power tracking and variations in DC input voltage. The *Design Handbook* presents three possible methods of battery interface, as shown in figure 7-2. The first shows the array and battery bank connected directly to the inverter. In the second, output from the array passes through a DC-to-DC converter on the way to the battery bank and inverter. The third schematic shows the battery bank separated from the array/inverter interface by a voltage regulator with a low-voltage disconnect to take the batteries out of the system when they reach the maximum recommended depth of discharge.

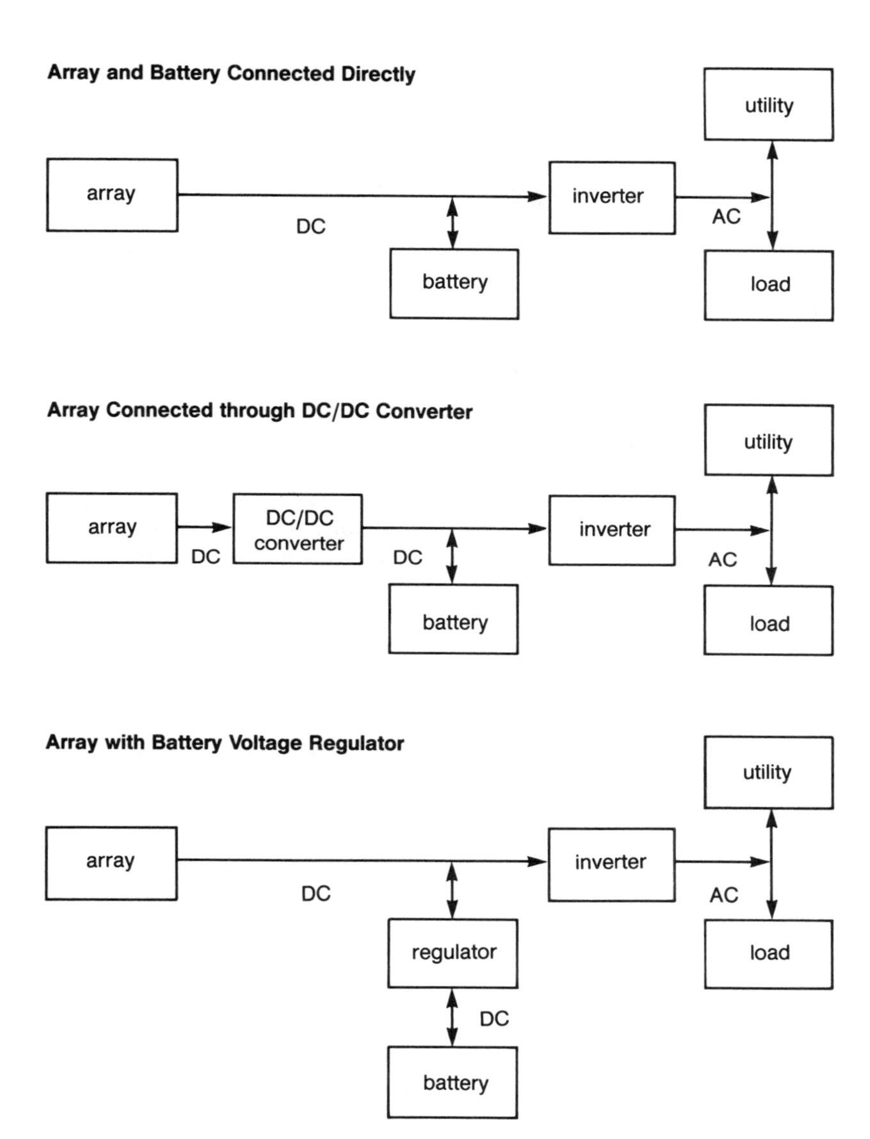

FIGURE 7-2
Block diagrams showing methods of storage battery interface in utility-interactive PV systems.

Case Studies: Designing Utility-Interactive Photovoltaic Systems

The remainder of this chapter will be devoted to a description of three UI photovoltaic systems that can serve as case studies in the techniques of system sizing and component selection. Unlike the hypothetical houses described earlier in this and previous chapters, these three examples are actual houses designed by my firm, Solar Design Associates. Two of them are located in Massachusetts and the third is in New Mexico.

The Southeastern Massachusetts House

Early in 1979, Solar Design Associates was commissioned to design a 2,500-square-foot, photovoltaic-powered, all-solar residence in southeastern Massachusetts. The clients were a semiretired professional man and his wife, a couple with a preference for comfortable surroundings and a commitment to high aesthetic standards, yet with a concern for energy conservation and self-reliance that not only enhanced their interest in solar energy but also led them to maintain a trim electrical load demand. As we shall see when we look at the details of PV array sizing for their house, this conservative approach to power use put them firmly in the low-demand category we defined in chapter 6 and resulted in array requirements that are the same as those outlined in our earlier hypothetical example.

The Southeastern Massachusetts house is a compact structure, built largely of masonry, concrete, and steel to assure a high internal thermal mass. The house is earth-sheltered on much of the north, east, and west sides, with exposed walls insulated to approximately R-40 through the use of the Dryvit system—6 inches of foam faced with a low-maintenance, cement-based, stuccolike material. The south-facing portion of the house incorporates full floor-to-ceiling glass for direct passive solar gain, which is augmented by an attached solar greenhouse.

Domestic hot water heating is provided by a 270-square-foot, flat-plate thermal solar collector array, which also serves to provide supplementary space heating. Most of the space heat not directly derived from passive solar gain is provided by burning wood. The house features two woodstoves, one in the main living area and one in the kitchen, which the owners also find convenient for some cooking tasks in the winter. An auxiliary electric furnace was installed for backup space heating but is rarely if ever used. During the summer months the house remains at or below 21°C (70°F) without the need for air conditioning.

With heating largely out of the picture and cooling a function of good ventilation and passive design, electrical loads in the house are principally the refrigerator, freezer, clothes dryer, and electric oven—along with lighting, occasional use of power tools in the workshop, and electronic entertainment equipment. The monthly load demand has, on occasion, dropped to below 300 kilowatt-hours, although 350 to 400 kilowatt-hours is closer to the norm. And there is nothing about the house to suggest austerity.

A utility-interactive photovoltaic system was designed for the house because the owners had opted against on-site battery storage, although ample space for a possible future battery bank was included as part of the design. Arrangements to connect with and feed power into the grid were satisfactorily concluded with Boston Edison Company, the local utility, and array sizing began.

The array size was influenced by both the load demand of the residence and the roof area available. The annual average daily peak sun hours of 3.8 kilowatt-hours per square meter for this southern New England loca-

PHOTO 7-2
The interior of the solar greenhouse at the Southeastern Massachusetts PV-powered house.

tion resulted in the design of a nominal 4.5-peak-kilowatt array. The modules chosen were Mobil Solar Energy Corporation's 40-watt Ra-40B, each with a rated output voltage (V_{no}) of 15.6 VDC and a nominal wattage of 40 peak watts at STC of 1 kW/m² and 25°C. One hundred and twelve modules were specified for the array, providing a nominal output of 4.48 peak kilowatts. Adjusted for the expected average cell temperature (NOCT) of 45°C, the actual output expected was calculated as follows:

$$(45°C - 25°C) \times 0.4\% = 8\% \text{ reduction for NOCT}$$

$$40 \text{ W}_p \times 0.92 = 36.8 \text{ W}_p\text{/module at NOCT of } 45°C$$

$$36.8 \text{ W}_p \times 112 \text{ modules} = 4{,}122 \text{ W}_p$$

$$4{,}122 \text{ W}_p \times 3.8 \text{ peak sun hr/avg day} = 15.66 \text{ kWh/avg day}$$

The examples presented earlier in this chapter demonstrate how to determine how many modules must be connected in series—once correction for NOCT influence upon output voltage has been made—in order to meet the input voltage needs of a utility-interactive DC-to-AC inverter. In

PHOTO 7-3
The south elevation of the Southeastern Massachusetts house, showing the roof-mounted solar thermal and photovoltaic arrays.

the case of the Southeastern Massachusetts house, the Windworks Company inverter that was chosen could have accepted an input voltage of 196.6 VDC, which would have been the result of stringing 14 of the Ra-40B modules (with a temperature-related reduction of voltage output to 14.04 VDC). Eight such strings in parallel connection would have been sufficient. However, the stronger voltage output of a 16-module series string was selected as being slightly more efficient with consideration for the inverter's input requirements. The result? A 16-series by 7-parallel configuration, handsomely mounted on the south-facing 50-degree roof.

The meters have been spinning at this house for over five years, and they spin a welcome story for its venturesome and pioneering owners. Over a two-month period in the fall of 1982, for instance, the total energy bill was $28, or $14 per month, to completely operate the house. And, tourist brochures aside, Massachusetts is not known for an unbroken progression of crystal-clear, sun-filled autumn days. When those days do come along, the clothes go on the line rather than in the electric dryer. That's all part of electrical husbandry and a modest effort to invest in order to enjoy this significant measure of energy independence.

The Santa Fe House

While the Southeastern Massachusetts house was being constructed, Solar Design Associates was commissioned to design a UI photovoltaic residence in Santa Fe, New Mexico. In early 1981, Rational Alternatives, a solar builder/developer based in Santa Fe, decided that the time was right to design and construct the nation's first privately funded speculative PV residence. Encouraged by a very positive reception from mortgage lenders, insurance underwriters, building officials, and the local utility, they retained Solar Design Associates to design the house. Ground was broken in the summer of 1981.

The 2,200-square-foot, single-story home features three bedrooms, two baths, a library, an open-plan living and dining area, a kitchen, a solar

PHOTO 7-4

The living room of the PV-powered residence in Santa Fe, New Mexico.

sunspace, an air-lock front entry foyer, utility and storage spaces, a two-car garage, an expansive landscaped south patio, and a walled private garden terrace on the north side of the house that includes a solar-heated outdoor hot tub.

The home is designed in the characteristic pueblo-style architecture typical of the Southwest. Exposed ceilings with vigas and pine decking, brick floors, and kiva fireplaces capture the charm of old Santa Fe. The house is earth-bermed on the north side and features walls of masonry and adobe to provide internal thermal mass with south-facing glass for direct passive solar gain.

A roof-mounted solar thermal collector array was included to provide heating for domestic water and the outdoor hot tub. Appliances are all electric; those chosen were the most energy-efficient models available on the market. Auxiliary space heating comes from wood burning, with electric resistance as a backup. Twenty homes of similar size and design built by Rational Alternatives were monitored by the local utility for a 12-month period during 1980 and 1981, and the results showed an average of $23 per year in space-heating costs.

The photovoltaic system for the Santa Fe house was designed to provide an average of 500 kilowatt-hours per month. The array was rack-mounted above the building's flat roof and features 78, 12-inch by 48-inch ARCO Solar 16-2300 modules. Each module consists of 35, 4-inch-diameter, single-crystal silicon cells in a single series string with redundant cell interconnects.

Each module produces 37 watts of power under STC of 1 kW/m^2 at 25°C. Adjusted to expected field conditions of the same insolation level but a NOCT of 47°C, each module has a rated V_{no} of 14.6 volts and a delivered power of 33.75 peak watts.

PHOTO 7-5
The south face of the photovoltaic-powered residence in Santa Fe.

Using these adjusted figures, the output of the Santa Fe house photovoltaic array was calculated as follows:

$$78 \text{ modules} \times 34\ W_p = 2{,}632\ W_p \text{ at NOCT}$$

$$2{,}632\ W_p \times 6.2 \text{ peak sun hr/avg day} = 16.32 \text{ kWh/avg day}$$

Thanks to the very favorable climate and resultant high insolation levels in Santa Fe, the 2.6-peak-kilowatt array has an average output of about 500 kilowatt-hours per month. Six strings of 13 series modules each are paralleled to provide the nominal 200-VDC input for the utility-interactive DC-to-AC inverter manufactured by the Windworks Company.

As with the first photovoltaic house we designed in Massachusetts, there is no on-site electrical storage at the Santa Fe house. All surplus power is sold to the utility and kilowatt-hours are purchased from the utility to make up any shortfall.

The Santa Fe photovoltaic house was completed in mid-February of 1982 and sold the first day it was offered on the market. The PV system produces over 6,000 kilowatt-hours annually, which, when taken together with the other solar and energy conservation aspects of the home, will likely make it energy-independent on an annual basis. The owners are delighted with the house because of the energy features and its Southwestern charm.

The Impact 2000 House

In early 1983, Boston Edison Company commissioned Solar Design Associates to design a state-of-the-art solar residence that would include a UI photovoltaic system. The house was constructed as part of the utility's Impact 2000 program to demonstrate trends in energy and housing design that will shape the twenty-first century. The home's design reflects the program's goals of energy conservation, load management, and renewable energy utilization.

PHOTO 7-6

The living room of the Impact 2000 house.

Located in Brookline, Massachusetts, about 8 miles west-southwest of Boston, the residence features 2,900 square feet of net living area, superinsulation, high internal thermal mass, and direct passive solar gain.

Appliances are all electric, including the high-efficiency, ground-coupled heat pump that provides space heating, air conditioning, and supplemental domestic water heating. Primary domestic water heating is supplied by a 125-square-foot solar thermal array integrated with the south roof.

The Impact 2000 house faces true south with the 576-square-foot photovoltaic array mounted on the 45-degree sloping south roof. Three hundred and sixty-five square feet of south-facing, double-glazed, heat-mirror glass allow passive solar gain to warm the living room, dining area, family room, sunspace, and bedrooms. The living room floor, which is dark quarry tile over a concrete base, helps store the passive solar gain.

A large masonry fireplace "core" in the center of the living room provides additional thermal mass and offers the option of wood heat. This internal core extends up through the house to the master bedroom on the uppermost level, where there is also a fireplace. Both fireplaces feature heat recirculation, have ducted-in outside air for combustion, and are fitted with glass doors. The master bedroom suite also features a separate dressing room, whirlpool tub room, three-fixture bathroom, and a private walk-out balcony. The kitchen, family room, and dining room are on the middle level, while the library, two more bedrooms, and another bathroom are located on the lower living level.

The outer walls have 6 inches of fiberglass batts plus 2 inches of rigid urethane under the vertical cedar siding (R-30), 18 inches of fiberglass insulation in the ceiling (R-60), and 8 inches under the floors (R-26). Double-glazed heat-mirror windows manufactured by Hurd are used on the east, west, and north walls to complete the low-heat-loss structure. One and one half stories of earth-berming on the north side and rigid insulation against the outer foundation walls also keep heat loss to a minimum.

A good portion of the space heating in this house will come from passive solar gain. Supplemental space heating is provided by the water-to-air, ground-coupled 48,500 Btu/hr heat pump with a low-pressure air circulation system for the entire house. Supply registers are in the floors near the outer walls, and air returns are at the top of the cathedral ceilings.

The low-speed fan may operate continuously to circulate the passively heated air throughout the house. The house is planned and designed so that natural-convection ventilation will reduce or eliminate the need for mechanical cooling. However, the heat pump can also be used to provide air conditioning if desired.

The photovoltaic system was sized to deliver approximately 15 kilowatt-hours per average day. This sizing decision was based principally on the availability of roof area and not from any in-depth analysis of the expected loads. The array features 24 large-area, 4-foot by 6-foot residential PV modules from Mobil Solar Energy Corporation. Each module is rated to produce 180 peak watts at STC and includes a positive and negative junction box termination and internal diode protection.

Adjusted to the 42°C NOCT expected at this site, each module can be expected to deliver a peak wattage of 167.76. The array output was thus figured as follows:

$$24 \text{ modules} \times 167.76 \text{ W}_p = 4{,}026 \text{ W}_p$$

$$4{,}026 \text{ W}_p \times 3.8 \text{ peak sun hr/avg day} = 15.3 \text{ kWh/avg day}$$

Quick-connect Solar-Lok connectors manufactured by AMP, Inc,. were used to wire the array in six strings of four 48-volt modules each to provide the nominal 200-VDC input voltage to the American Power Conversion Corporation Sunsine utility-interactive inverter.

PHOTO 7-7

The south face of the Impact 2000 house.

The south roof was divided into two sections for the photovoltaic array, each with 12 modules (two rows high by six wide). The middle section of the roof between the two PV subarrays was reserved for the roof-integrated solar thermal collectors and the sloped glazing for the sunspace. The PV array was also roof-integrated, with the large-area Mobil PV modules replacing the conventional roof sheathing, building paper, and roofing to become the structure, weatherskin, and finished roof.

The details of the roof-integrated photovoltaic array installation as well as all the other aspects of the construction of the Impact 2000 House were chronicled step by step for ten million television viewers on a 26-week national television series on the PBS network. The series, "This Old House," had up until then featured only renovation work. For this occasion it was renamed "The All-New This Old House" and the 1983–1984 season was the most popular in the show's five-year history.

Boston Edison was pleased with the house as well. "The Impact 2000 house incorporates many state-of-the-art energy features with a contemporary design that has wide public appeal," says Edison's project director Carl Gustin. "The house won both state and national awards for innovation and over 10,000 people visited it during our public tour at the conclusion of the television series."

At the beginning of this chapter we pointed out some of the compelling arguments that have been raised against total dependence on a highly centralized power-generating system. Among the major problems inherent in a centralized system are the tremendous capital outlays that it demands, the negative environmental impact produced by the type of technology that it employs, and its vulnerability to sabotage or breakdowns caused by human error.

Despite the validity of these criticisms, we have to admit that the breakup of the conventional utility grid currently in place in this country would be both undesirable and impractical. Even the most ardent advocates of energy conservation and decentralization are forced to agree that the benefits so far granted by our utility infrastructure must be preserved, both to maintain the convenience on which so many have come to depend and to continue the promise of prosperity and social betterment for those who may not yet be so comfortable.

Therefore, lying behind the discussion in this chapter is the assumption that, although the increased development of effective on-site means of generating electric power over the next years and decades will *not* lead to the demise of the electric utility companies we have today, it is naive to expect that business will continue as usual for these firms.

What *will* take place might best be described as a transformation, or evolution, of these companies and their traditional ways of doing business. Rather than functioning as exclusive producers and suppliers of electricity, they will assume a greater role as distributors of power generated at many different places. They will, in effect, become "energy banks," receiving deposits of electrical power from a dispersed network of small-scale producers while simultaneously feeding power back out onto the grid to consumers who wish to purchase it.

To be sure, the utility companies will continue to produce electrical power for the foreseeable future. But their output capacity in the future will not need to be as extensive or as centralized as it is at present, since it will be supplemented by power purchased from small-scale producers. Moreover, a substantial part of the power produced by the big companies in the future will no doubt originate from large, central-station photovoltaic array fields, rather than solely from generators driven by fossil fuels.

The case studies in this chapter have shown that a fully equipped contemporary house with a 4.5-peak-kilowatt photovoltaic system can get by with only $28 worth of grid input during two months of a New England autumn; that a Southwestern 3-peak-kilowatt PV residence can produce 6,000 kilowatt-hours of electricity a year, leading to virtual energy independence; and that a Massachusetts utility company was sufficiently impressed with PV technology to sponsor the building of a 4.5-peak-kilowatt, 2,900-square-foot luxury home as a demonstration of the promise inherent in the cooperation between central grid suppliers and dispersed residential PV generation.

Each time a utility-interactive PV installation is completed, component manufacturers, system designers, and the interconnecting utilities learn more about the efficient generation and use of dispersed, line-tied PV power. The potential for decentralization is growing, and it stands to benefit utilities and consumers alike. But line-tied photovoltaics is just part of the solar electric story. In the next chapter, we'll look at how a stand-alone PV system is put together and what it is like to live with one.

CHAPTER EIGHT

STAND-ALONE PHOTOVOLTAIC SYSTEMS

Stand-alone (SA) photovoltaic (PV) systems represent a vast and varied range of possibilities. With the exception of sunlight itself, what could be more versatile than site-generated, no-strings-attached electricity? A book this size can hardly begin to catalog all the uses for SA photovoltaics, let alone describe them in detail. Independent arrays power irrigation pumps and illuminate villages in the Third World. They also run cathodic protection systems and remote telecommunications facilities both here in the United States and around the globe.

Even before photovoltaics began to make inroads as a source of electricity for full-size, year-round homes, many thousands of SA installations made the comforts of modern life possible in small, remote homes and vacation cabins throughout the world. And, although these early applications were mostly small, industry has now picked up the concept of SA photovoltaics.

Appropriately enough, it is a photovoltaic module manufacturer that has created one of the most spectacular examples to date of a solar-powered, energy-independent manufacturing facility. The Solarex Corporation's PV "breeder" plant, located in Frederick, Maryland, depends solely on its roof-mounted array for 200 peak kilowatts of electrical output to provide all the energy required for heating and cooling, as well as the power needed to operate office equipment and production machinery. The PV modules built at the plant are thus themselves the product of solar electricity, with no assistance furnished by fossil fuels or the utility grid.

A stand-alone solar electric system presents its users with a challenge that is both technological and philosophical. It asks them to use not only their ingenuity but their common sense as well. The SA solar electric house, like a space station, is a self-contained unit—at least as far as kilowatt-hours are concerned.

Bringing Down Load Requirements

At this point, anyone seriously considering an SA photovoltaic system should go back and thoroughly review chapter 6, since the ideas on loads, peaks, and conservation presented there apply with special emphasis to homes powered by stand-alone PV systems. The most important theme that emerged from our examination of the electrical loads in three hypothetical homes was that electrical devices designed to generate heat are responsible for the largest proportional draw of current in the home.

Reliance upon electric furnaces or baseboard resistance units as primary means of space heating, already a poor economic proposition in all but a very few parts of the country, is definitely not a good idea with photovoltaics—even in an up-scale, no-holds-barred project like the Carlisle house. This is especially true for SA systems. Nor do electric clothes dryers, ranges, or hot water heaters make much sense. It is far better, considering both economic and thermal efficiency, to use natural gas or bottled propane for cooking and solar thermal collectors for domestic water heating.

Begin planning well in advance for your SA photovoltaic system by plotting out your loads carefully and trimming where you can. Plan to replace older, less efficient major appliances with new, energy-efficient, state-of-the-art models. Consider the appliance replacements to be part of the initial system outlay, to be amortized along with the rest of the hardware over the life of the house. It will almost certainly be less costly to purchase new energy-efficient appliances than to install additional PV array capacity to power your old, inefficient ones.

Once you have arrived at an estimate of total load demand for your SA photovoltaic installation, the next step is to calculate the physical size and rated output of the PV array required to handle it. As you begin this step of the process, it is important to remember that your load requirement figures are not written in stone: if economics or prudent system design suggests a reduction, there is nearly always likely to be some extent to which the monthly kilowatt-hour tally can be pared down even further, without seriously compromising comfort, convenience, and lifestyle.

Remember also that beyond the question of basic kilowatt-hour requirement is the question of when individual loads present themselves and what their contribution is to the total peak demand. These are factors that influence not only the size of a PV array but also the extent of storage capacity and the size of whatever backup facilities might be envisioned.

Array Sizing for Stand-Alone Systems

Most stand-alone owners are not satisfied to have their auxiliary generators cut in with the same regularity with which grid power enters a UI photovoltaic residence. There are several obvious reasons for this: generators use fuel and require maintenance, both of which cost money; generators are noisy and intrusive, especially at night; and generators do not buy back excess power put out by the array. Thus, the sizing of an SA system must be more precise than the sizing of a line-tied system. The SA system must provide enough reserve power in the batteries to cover peaks, evening use, and periods of bad weather to the greatest extent possible.

With this in mind, let's add some important additional steps to the sizing equations presented in previous chapters. These steps cover the losses in the battery bank, inverter, and other system components. The largest losses are battery related, since, as we learned in chapter 4, the round-trip efficiency of a battery bank in good condition is about 80 percent.

We also learned in chapter 4 that if sizing calculations are done in amp-hours (Ah), the battery losses do not have to be figured separately since they are compensated for by the fact that the actual operating voltage of most photovoltaic modules is higher than the nominal voltage rating (usually 16 to 18 VDC for a 12-volt nominal module). In all but the hottest climates, this extra margin of voltage is enough to satisfy the battery losses. Thus, when amp-hours are used, the system designer usually does not have to calculate the effect of temperature on module output.

However, in very warm locations where average daytime outdoor temperatures will exceed 35°C (95°F) for significant periods, the effects of temperature must be considered. In response to the impact of differing temperature conditions on array operation, ARCO Solar has begun to market three different 12-volt (nominal) photovoltaic modules, each with a different number of cells in series (30, 33, and 36, respectively). The selection of the appropriate module is important, since not only is it possible to lack sufficient voltage to fully charge a battery bank, it is also possible to have too much voltage.

For example, if a 36-cell module were to be specified for a battery-charging application in a temperate climate, the battery bank voltage would determine the voltage at which the photovoltaic array operated and would drive the high array voltage down and away from its maximum power point. Since PV modules are sold by the watt, the customer would have purchased unusable power. (Figure 8-1 compares the voltage output of these three modules at two different operating temperatures.)

In demonstrating the sizing process for an SA system, we will base our array-sizing calculations on the amp-hour method. However, if we were to calculate the array sizing in terms of watt-hours, the figure on which we would base the amount of battery loss is only that portion of the estimated daily load that is not supplied directly by the array. This will vary according to the load profiles of individual homes, but on the average, 35 percent of the daily electrical load for a typical SA residence will be delivered during the day directly from the array and the remaining 65 percent will be delivered via the battery bank.

For the sake of simplicity, let's assume that the outdoor ambient temperature at our hypothetical site will not rise above 35°C except for a few summer afternoons. Let's also assume that we are dealing with a rather

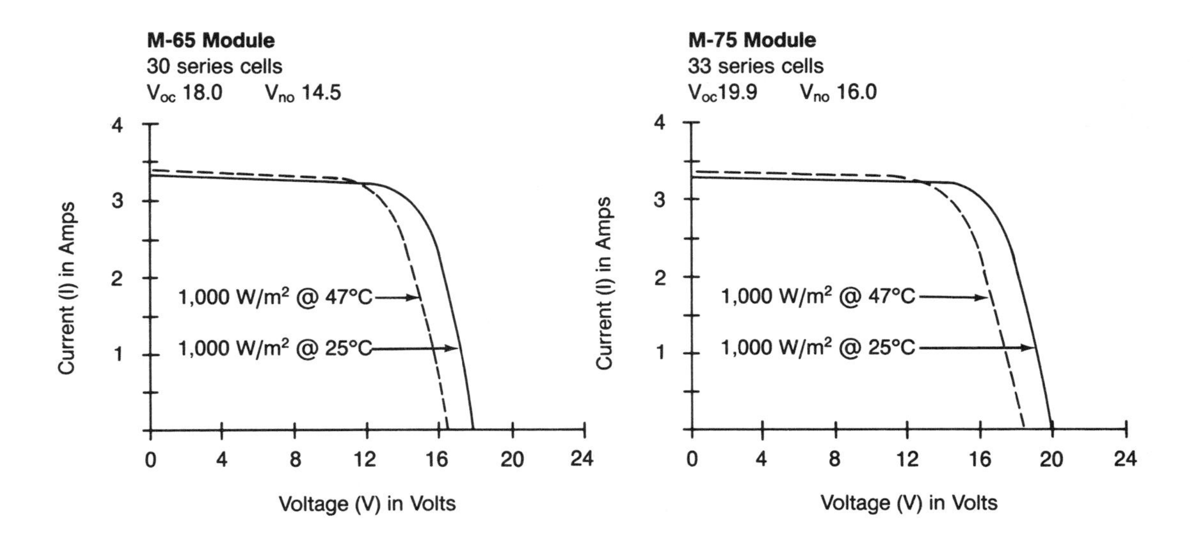

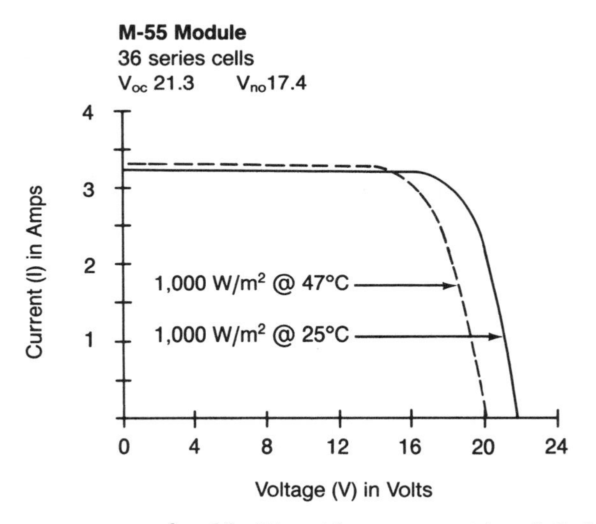

FIGURE 8-1
I-V curves for three ARCO Solar modules, showing voltage variations caused by change in operating temperature from 25°C to 47°C.

comfortable SA residence, one with a daily load requirement of 7.5 kilowatt-hours.

After the basic (net) daily load demand has been established, we must add the losses from the inverter and the other components (excluding batteries) in the system. If we choose a good-quality, high-efficiency inverter, we might obtain an average efficiency of 90 percent. If we assume that 80 percent of this SA system's load demand of 7,500 watt-hours (Wh) will be delivered through the inverter, our figures look like this:

7,500 Wh × 0.8 = 6,000 Wh/day delivered through inverter

6,000 Wh × 0.1 = 600 Wh/day lost through inverter

Next we calculate the internal system losses from components other than the battery bank and inverter. These losses occur in the voltage regulator, the wire runs, module mismatch, diodes, and the like. In chapter 3, we established as a rough rule of thumb that we can add 10 percent of the daily load for these internal system losses. Careful system design and selection of quality, high-efficiency components could reduce this number to 5 percent or even less for a system of this size. For this example, we will use a figure of 7.5 percent. We calculate:

$$7{,}500 \times 0.075 = 562.5 \text{ Wh/day additional system losses}$$

We can now add our inverter and internal system losses to the base load to find our total daily requirement:

$$7{,}500 \text{ Wh} + 600 \text{ Wh} + 563 \text{ Wh} = 8{,}663 \text{ Wh}$$

We conclude that we need to derive 8,663 watt-hours of electricity from the array per average day.

Consulting a source on insolation data, we find that our hypothetical site has an average winter peak sun hour figure of 3.8. With this information we can now calculate the number of peak watts (W_p) needed in the system by dividing the total number of watt-hours needed by the average daily insolation available:

$$8{,}663 \text{ Wh/day} \div 3.8 \text{ peak sun hr/day} = 2{,}279.7 \text{ W}_p$$

We conclude from this exercise that the array wattage requirement is 2,280 peak watts. Note that we used the average winter peak sun hours for this array-sizing example rather then the annual average peak sun hours employed in some earlier examples. The reason for this is that winter is usually the time of both greatest load demand and lowest available sun. With a stand-alone system, you should size for the worst combination of load and insolation.

Here is where the array-sizing method for SA systems, which is based on amp-hour calculations, departs from the UI examples based on watt-hours presented in the last chapter. We must convert the load to amp-hours before sizing the array, but first we must select the system's DC voltage. This decision will determine the operating voltage of the inverter, the battery bank, the regulator, the array, the battery charger (if one is included), and any DC appliances that are to be served directly from the battery bank.

Of all these voltage considerations, that of the inverter is probably the most important. Fortunately, the inverters designed for SA service offer somewhat more latitude in the choice of DC-input voltage than their line-tied counterparts. Units are available in 12-, 24-, 48-, and 120-VDC input, as well as a number of other less popular ratings such as 32 and 36 volts.

In a system of this size, I would specify an inverter with a high DC input voltage because of the efficiency considerations discussed in chapter 5. Our options are 48 or 120 VDC. If we select a 48-VDC inverter, we then divide our array wattage figure of 2,280 by 48 and learn that our system must be able to produce 47.5 amps at maximum power output.

Let's use the same 40 peak watts (36.8 W_p at NOCT) modules we used in the case study of the UI photovoltaic system at the Southeastern Massachusetts house for this array. We find from the manufacturer's literature that the module will deliver 2.48 amps under full sun. We know from chapter 3 that we must use the module's nominal voltage rating when sizing an array using amp-hours and that we will require four 12-volt (nominal) modules in series to form one 48-volt (nominal) array string. To determine the number of module strings required for our array, we must now divide the load in amps per hour (47.5) by the module output in amps per hour (2.48). From this we learn we need 19.15 series strings. Since we cannot size in a fraction of a string, we will round up to 20 series strings of four 12-volt (nominal) modules each.

This produces an array of 80 (4 series by 20 parallel) 12-volt (nominal) 40-watt modules, which will generate a daily surplus of about 381 watt-hours or a little more than 8 amp-hours (at 48 VDC nominal) more than our total daily load requirement of 8,663 watt-hours. And that's good. Remember, this is a stand-alone system; as such, it must satisfy the load or fall back on the output of an auxiliary generator. (For the sake of this example, we have assumed the generator will play only a standby role in the system.) Remember too that loads are more likely to increase than decrease over time, especially if conservation efforts have been carefully implemented in the initial design of the house and its power system.

It should be pointed out that if there is no generator included to provide backup power to a stand-alone photovoltaic system, then the array needs to be sized somewhat (approximately 10 percent) larger to provide an additional margin of capacity. This will ensure that after a prolonged period of no sun the load will be served and the battery bank will be recharged without an interruption in service.

The 80-module array that we have settled on should satisfy our requirements very nicely if we have a properly sized battery storage bank. So let's review the battery storage sizing technique first presented in chapter 4, using this system as a working example.

Storage Sizing

Determining the extent of battery storage that is required in an SA photovoltaic system is every bit as important as sizing the array itself. Batteries are what make SA photovoltaics flexible; they are what enable us to reconcile daytime's sunshine harvest with evening's proportionally heavy load demands. Batteries must provide more than just enough margin of electrical storage to carry us through the night. With the knowledge from local weather data that there will be occasional prolonged stretches of inclement weather that will severely reduce insolation, the storage system should be capable of providing backup power for several days. A minimum of two days' storage without reliance on nonsolar auxiliary generating equipment is recommended; three to six days' storage is common.

Overall load demand, peaks, and desired length of coverage (system operation over days of no sun) are all determining factors in the sizing of a storage bank. Another is the durability of the batteries themselves: the deeper a battery is repeatedly discharged, the shorter its life span becomes. This fundamental principle is true even of the "deep-discharge" batteries designed especially for photovoltaic systems. Thus it may make sense to use a larger number of batteries and to operate them at a slower charge/discharge cycle rate and a shallower depth of discharge (DOD).

Once the basic decisions have been made about the type and capacity of the individual batteries that will make up the storage system and their recommended daily depth of discharge (DDOD), the rest of the sizing procedure becomes a matter of basic arithmetic. Let's review this procedure using the loads from our array-sizing example. Let's assume that we want a storage capacity of three full days without drawing the batteries past their recommended DDOD and the possibility of going seven days with no sun by either deeply discharging the batteries or substantially reducing the loads. (In a worst-case scenario we could, of course, make things easier on the batteries by turning on the auxiliary generator.)

We previously calculated our daily load (including internal system losses) to be 8,663 watt-hours. At 48 volts this translates into roughly 180 amp-hours per day (8,663 ÷ 48 = 180.47). Over three days, that amounts to a total requirement of 540 amp-hours of available storage. Over seven days the total is 1,260 amp-hours.

For this example, let's choose to use a battery such as the Surrette model T-12-140, which is rated to deliver 12 volts and 140 amp-hours when discharged over a 20-hour period and is designed for deep-cycle service. Connecting four of these batteries in series meets our design goal of 48 VDC. We now have to determine how many of these series strings will provide us with an aggregate available battery capacity of 1,260 amp-hours to deliver seven days of storage, assuming a maximum recommended depth of discharge (MDOD) of 80 percent.

Since each battery has a total rated capacity (C) of 140 amp-hours, this translates into a total usable capacity (TUC) of 112 amp-hours at an MDOD of 80 percent. Dividing our storage TUC requirement of 1,260 amp-hours by 112, we find that we need 11.25 of these series strings. We round this figure up and specify a battery storage bank of 48 batteries in 12 series strings of 4 batteries each, as shown in figure 8-2.

How does this stand up against the kilowatt-hour requirements we have established for our house? At 8,663 watt-hours per day, three days' full load equals 25.99 kilowatt-hours and seven days' full load equals 60.64 kilowatt-hours. At 80 percent depth of discharge, each series string of batteries can deliver a total of 5.38 kilowatt-hours (48 V × 112 Ah = 5,376 Wh) and the entire storage bank can deliver 64.5 kilowatt-hours (5.376 kWh × 12 strings). That more than covers the full load over a period of seven days.

The total rated capacity of this battery bank is 80.64 kilowatt-hours (48 V × 140 Ah × 12 strings). If we assume that the "average normal" demand on the storage bank will be three days of load, we find that we are

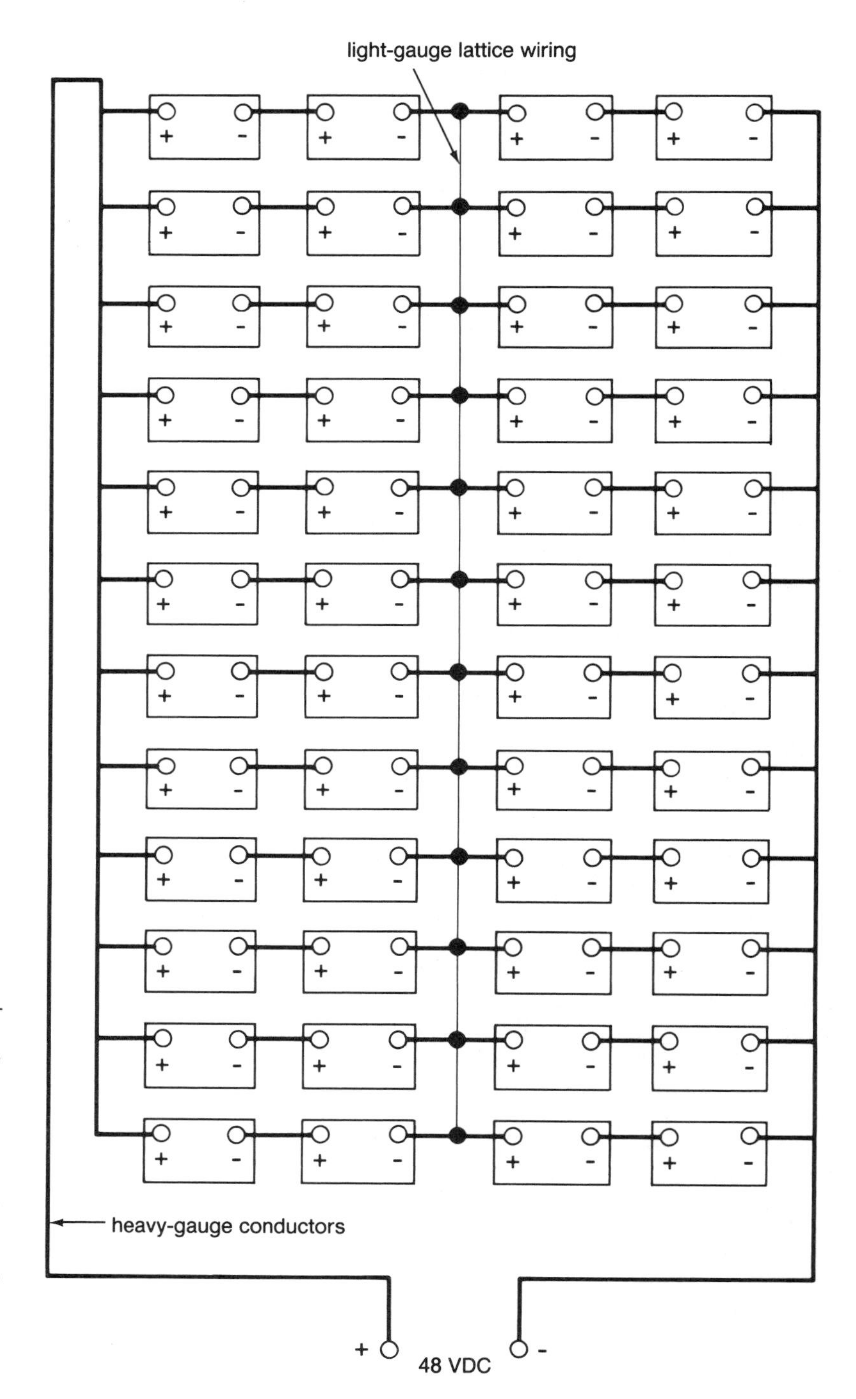

FIGURE 8-2

A schematic wiring diagram of a battery storage bank showing 48, 12-volt batteries, each with a total usable capacity (TUC) of 112 amp-hours, wired in 12 series strings of 4 batteries each to deliver a TUC of 1,344 amp-hours at 48 VDC nominal. (Note light-gauge "lattice" wiring at center of strings.)

cycling to a DOD of about 32 percent (25.99 ÷ 80.64 = 0.322) over each three-day period and to about 11 percent (8.66 ÷ 80.64 = 0.107) on a daily basis, which is very good.

This approach to battery storage system design is rather conservative. It involves rather high initial costs, but it offers excellent performance over the long term by greatly extending the life of the batteries. However, we could move in the other direction and reduce both the size of the battery bank and the system cost while still obtaining long battery life by relying on the auxiliary generator after the third day of no sun instead of the seventh. If this were done, the recommended sizing for the battery bank would be five series strings of four batteries each.

Backup System Options

Integrating a photovoltaic array with other on-site power-generating equipment provides assurance that batteries will be kept charged despite the vagaries of natural forces. For all practical purposes, the only viable alternative power sources are generators powered by wind, hydropower, or fossil fuels. Let's take a brief look at each of these options.

Like photovoltaic modules, most wind generators put out DC electricity. They can thus provide a direct charge to a battery bank, assuming that the battery voltage agrees with that of the generator. The batteries used with a wind system should be the same deep-discharge type as those recommended for photovoltaics, which means that both input sources can charge the same batteries. After proper battery selection, completion of the PV/wind system becomes a matter of selecting the proper balance-of-system components. For the most part, this should be done following the same criteria that would apply to an all-PV system.

The wind/photovoltaic combination sounds ideal. Indeed, wind power is very cost-effective at the right site. When the sun doesn't shine, the wind will probably blow. However, it seems that the highest sustained winds most often occur at sites where people don't choose to live. This makes wind a perfect candidate for central-utility generation but not necessarily the best system for the individual homeowner. Despite the number of times residential clients have approached my firm expressing the desire to install a wind turbine or a PV/wind hybrid system, we have yet to find a site with enough average wind speed to justify the investment.

Hydropower is an even more site-specific resource than wind, depending as it does on a flowing watercourse with sufficient head to make a turbine operable. Water as a power source offers the advantage of relative constancy and predictability. Constancy does not count for everything, however, and a persistent problem with small hydro systems is that they are often unable to cover daily peak loads. And if there is enough water power available on a round-the-clock basis to charge an adequately sized bank of batteries, why bother with a photovoltaic system at all?

This brings us back to the "old standby," the diesel- or propane-fired generator. Anyone who has spent much time in off-grid areas will be familiar with generators, if for no other reason than the noise they make during operation. But noise and exhaust are only part of the inherent prob-

PHOTO 8-1
In certain installations, a wind turbine such as the one shown here can work well in conjunction with a photovoltaic array.

lem in running a generator fueled on hydrocarbons for any substantial length of time. The price and availability of fuel has to be considered, along with the substantial expense of the machines themselves. And of course, the longer and more often you run a generator, the quicker it wears out. Thus, the cost of regularly replacing parts and entire units has to be figured into the price of this method of producing electrical power.

As we saw in chapter 2, costs per kilowatt-hour rise substantially when a generator is used below its rated capacity. It makes sense, then, to use the unit for as brief a period and at as full a load as possible. There are two ways to accomplish this: (1) carefully concentrate loads into one part of the day (practical for some loads, impractical for others); or (2) use the generator to charge batteries, which in turn can meter out the power available to loads over a longer period.

If the power system's batteries are normally charged with solar electricity rather than by the generator, the burden on the generator is greatly reduced. This is preferable anyway, since high power output from the generator is not what batteries normally need. As was pointed out in the performance criteria provided in chapter 4, a battery's ability to receive and store charge is directly related to the *rate* at which it is charged, in terms of amps per hour. This is a job that is perfectly suited to photovoltaics, since it provides a moderate charge over many hours.

The diesel- or propane-driven generator is thus commonly used with SA photovoltaics in a strictly backup function, serving only to see the system through prolonged bad weather and to give the batteries an occasional equalizing charge. As can be seen in our case studies, with the right design, the PV/generator combination can produce a very good remote power system.

Stand-Alone Case Studies

Perhaps the greatest virtue of an SA residential photovoltaic system is its flexibility. The two primary ingredients—PV panels and batteries—are completely modular and thus may be configured in response to a broad range of load requirements. Choices exist between alternating and direct current and between high and low voltage. Hybrid systems combining AC and DC service can be created for maximum efficiency. The inverters, regulators, battery chargers, and auxiliary generators that make up the balance of the installation are all available in many sizes and power ratings, so there is no need for adapting components ill-suited to the desired application.

Best of all, SA photovoltaics leaves us with tremendous geographical leeway in choosing where we want to live, since it frees us from the notion of homesite selection based upon the availability of utility-grid power. The adaptability of SA systems is demonstrated graphically in the following examples of owners of PV-powered homes who have managed to satisfy their lifestyle requirements without any reliance on the power grid.

The Bleicken Residence

Kurt Bleicken goes to work every morning in southern New Hampshire, eastern Massachusetts, and South Carolina. Putting into practice futurist Alvin Toffler's "electronic cottage" idea, Kurt keeps on top of his business affairs—in this case the raising and management of capital for corporations and partnerships in business and real estate—through the use of his personal computer and a telephone modem. The computer, like all the other electrical loads in Kurt and Janet Bleicken's charming country home, is powered by their SA photovoltaic system.

When the Bleickens decided to build their home in rural New Hampshire they said "no, thanks" to the local utility's offer to extend grid service to their ridgetop homesite, which enjoys an expansive view of the valley below. The Bleickens have 21 acres, and Kurt estimates that the per-acre price he paid was only about a fifth of what property served by the grid would have cost in the same area. His entire photovoltaic system cost less than the $15,000 the utility wanted to charge for a grid hookup. With no monthly utility bills to follow, the PV choice was an easy one to make.

The Bleickens' photovoltaic system is a modest one. It features 16 43-watt ARCO Solar M-53 modules, a deep-cycle battery bank, and a high-efficiency inverter to deliver 120 VAC. While using far less electricity than the owners of a comparable-sized house tied to the power grid, they enjoy the benefits of not only a computer but also a conventional assortment of appliances, lighting, television, stereo, power tools, and other household equipment. To reduce their electrical load, they use propane for cooking, refrigeration, and clothes drying. Their well water is delivered by a DC-powered submersible pump made by Jacuzzi, which reduces inverter losses on their system.

The Bleickens have been living with photovoltaics for several years now, and the only problem Kurt reports with the system is the need for a

PHOTO 8-2

Jan and Kurt Bleicken's stand-alone PV-residence near Peterborough, New Hampshire, with the array mounted on the garden shed.

power supply filter on his computer to tame the fluctuations in voltage output from the inverter. Jan, an artist who paints large-format landscapes and nature scenes, will soon add an electric elevator lift in her studio so that she can get full and easy access to her large canvases.

During the last week of September 1985, when hurricane Gloria rocked New England, the Bleickens were the only people in their area who enjoyed uninterrupted electricity and all the comforts of home. "It sure turned some of the skeptics around," reports Kurt. "The eye of the storm passed right through our town and power was out for several days in some areas. We never had so many friends come to visit."

The Rolfson Residence

The Bleickens designed and built their country home themselves, but there are many locations in which fine, old houses stand off the grid, ready to be brought comfortably into the twentieth century with SA photovoltaics. Near the village of Albion, Maine, Eric and Becky Rolfson renovated a lovely 200-year-old Cape Cod that has never had a utility connection. "The power company wanted over $10,000 to extend the lines roughly 3,000 yards to our house," says Eric, who works in the development office of nearby Colby College. "For about half of that after tax considerations, we installed an eight-module PV system that gives us more power than we can use."

The Rolfsons have 12-volt service for their DC radio, television, and stereo; 24 VDC for lighting; and an inverter to provide 120 VAC when they need it. "We also have a propane generator, which we use occasionally between late November and February," Eric adds. "By April, it hardly ever comes on and by late spring, we're actually spilling power."

Stand-alone photovoltaic success stories like the Bleickens' and the Rolfsons' have become almost commonplace over the past several years as more and more people in all parts of the world come to fully appreciate the versatility and reliability of photovoltaics and the new degree of freedom that the resulting energy independence provides.

The Higgins Residence

On the coast of Labrador in Canada is a remote fishing village called Paradise River, which has no roads, no cars, and no utility grid. My firm designed a modest ten-module system for Bart and Charlene Higgins to power their cabin there. This was one of our most unusual projects, and in a way it was one of the most successful, not only because the system has performed beyond expectations but also because it lays to rest the notion that photovoltaics is a Sunbelt technology. Anyone who has ever been to Labrador knows that the area presents what must be close to the ultimate test for SA photovoltaics. Happily, it passed the test.

The Higgenses came to Paradise River from Toronto, where Bart had been a professor of Canadian literature and history and Charlene had been a systems analyst. Several years ago, in a radical career and lifestyle change, they moved to the tiny coastal village of 100 people and began making their living from Labrador's rich offshore stocks of salmon and cod. When they moved into their log house, they had no electricity, so they installed a gasoline-driven generator to run their washing machine and recharge their battery-operated radio. For lighting, they used kerosene.

In the spring of 1984 we completed the design for the Higginses' photovoltaic system, and the components began arriving at Paradise River by boat when the ice pack opened up that summer. Bart Higgins installed the 400-watt, 12-volt system himself, mounting the modules on a wooden support stand above the cabin roof and installing the deep-cycle batteries, regulator, and inverter next to his generator. By the time the long Labrador winter had again closed in, solar electricity had freed the family from its dependence upon the gas generator and kerosene lamps. "The biggest impact the PV system has made on our lives is in the amount of usable time we have each day," says Charlene. "It's expanded from about 7 or 8 hours in the winter to about 14 hours."

"Even in the dead of winter, the system has kept up with our electrical demand," says Bart. "And in late April, we're at full charge by 9:00 or 10:00 in the morning." In addition to their lights, the Higginses run two television sets, a computer, a short-wave radio and a regular radio, and several appliances, including a sewing machine. Bart plans to add a water pump and a small freezer in the near future. "At the end of February, the system's output begins to exceed our demand. We're thinking of getting a color TV now to help use up our surplus electricity."

The McCluskey Residence

Bart and Charlene Higgins chose SA photovoltaics for the simple reason that there was no utility with which to connect. By contrast, Don and Dorothy McCluskey had the option of connecting to the grid when they built their summer home on Block Island off the coast of Rhode Island in 1979. However, like the Bleickens and the Rolfsons, they were faced with a high connection cost. "The utility wanted over $15,000 for the connection and their rates were over 30 cents per kilowatt-hour," says Don, who is an electrical engineer.

Faced with those economics, the choice was clear. "We just couldn't justify connecting to the utility company," says Dorothy, a four-

PHOTO 8-3

Don and Dorothy McCluskey's comfortable vacation home on Block Island off the coast of Rhode Island receives all its power from a stand-alone PV system.

term Connecticut state legislator active in energy and environmental issues. "So we designed our house for energy efficiency and installed our own diesel generator. But we weren't happy with the generator. We came to Block Island to relax and enjoy its beauty and the peace and quiet. The generator was very noisy and required a great deal of maintenance."

In 1981, Don designed and installed a four-panel photovoltaic system with a bank of deep-cycle storage batteries, a voltage regulator, and an inverter. "I was really pleased," he says. "We went from 3,000 hours of generator time per season to about 300. This dramatically extended the life of our generator while significantly reducing maintenance, and it gave us two sources of power so we would never be without. Most important, it gave us the quiet we came to the island for." The McCluskeys have since increased their PV array to 800 peak watts and have disconnected and sold their generator.

Sunrise Technologies

So pleased was Don McCluskey with his photovoltaic system that he decided to share his knowledge and newfound quiet and independence with other people on the island. In June 1983, he and close friend Nancy Greenaway started Sunrise Technologies to design and install PV systems. Nancy knew that Kim Gaffett, the owner of Mid Ocean Press, was constructing a new building to house her printing business. She and Don approached Kim with a novel idea: why not power her new building with solar electricity?

Kim was intrigued but needed convincing that the solar array would really provide enough power to run her presses. Don invited Kim to see the photovoltaic system at his house and did some engineering calculations for her, and the deal was struck. Kim would allow the installation of a

PHOTO 8-4

The 1-kilowatt roof-integrated array atop Sunrise Technologies' building on Block Island.

1-kilowatt PV system on her roof, Sunrise Technologies would lease space in Kim's new building and provide themselves and Mid Ocean Press with electricity.

"The system works flawlessly," says Nancy, "Kim and I have all the power we can use." Kim is pleased, too. "When the local power company goes down, which happens at least eight or ten times a season, we're the only shop open and people are really amazed," she reports. All of Kim's work now bears her new trademark: Printed with Solar Electricity.

The Bournazos Residence

When Charlie Bournazos first saw the house on the point at the end of Duxbury beach in Plymouth, Massachusetts, it was a simple summer cottage with a small gasoline-driven generator. He liked the location so much he decided to buy it, increase the living space, and completely renovate and winterize the structure. But what about power? The generator was unreliable and expensive to run, and the noise it created greatly disturbed the peace and quiet that made the site so attractive in the first place. Although his house lies within sight of the Pilgrim nuclear power plant on the Plymouth shoreline, he could not get utility service.

About this time, Charlie overheard some people discussing photovoltaics over dinner in his restaurant, and he made some inquiries as to how practical it might be to power his home with such a system. The site proved to be perfect for PV power, since it featured plenty of south-facing roof, a modest electrical demand, and an existing generator in place for backup.

Charlie installed a 700-peak-watt PV array; a 24-volt, 2,500-watt inverter; a deep-cycle battery bank to provide three days' storage; and fully automatic controls and transfer switching. To provide telephone service, he bought a cellular telephone, which can operate either in his house on a DC power supply or in his car when he's away. He can call anywhere in the world with his hand-set and the small portable transceiver. Charlie reports that his PV system has performed flawlessly. His generator running time in the winter is very low, and in the summer the generator is turned on only to exercise it once a month.

At first glance, there may seem to be more differences than similarities among these SA photovoltaic pioneers—between the lives of the manager of venture capital and the printer, the engineer and the fisherman, the college fund-raiser and the restaurant owner. But they all have one thing in common, as do the tens of thousands of people worldwide who live with stand-alone PV systems: an enthusiasm for the decision they've made and the way it's working for them.

"We've found that even with a modest system, we can generate all the electricity we need," says Kurt Bleicken. "It's really changed our lives," says Charlene Higgins, "and I expect some of the other people in our town will soon follow our lead." Says Eric Rolfson, "It's really the ticket—we're absolutely delighted."

The beauty of SA photovoltaics, aside from the basic elegance of the technology, is in its flexibility and the options this creates. With a stand-alone system, you can select where you want to live—without regard for the utility grid and often at substantial savings in real estate costs—and choose what kind of life you wish to lead. Whether commercial fishing or high finance, climate and technology pose no limits.

Moreover, SA photovoltaics need not be limited to small, remote dwellings such as the Higgins cabin in Labrador. Quite the opposite is true. My firm recently designed a 12,000-square-foot, energy-independent residence for a California businessman. The mountaintop retreat is to include an indoor racquetball court, an Olympic-size swimming pool, and virtually every electrical convenience possible—from a high-frequency-link telecommunications system, an irrigation system for the orchard, and a satellite receiver dish, down to the landing lights on the private airstrip—all powered with photovoltaics.

The Maine Island House

Let's turn now to another example of a stand-alone residential system and examine it in greater detail, reviewing the design calculations required to provide a steady supply of electricity in a place where grid power is decidedly *not* available—a small island off the coast of Maine.

Between Eastport and Block Island, there are over 2,000 islands located off the New England coast, the majority of them off the coast of Maine. Some are just an acre or two in size, while others cover many square miles. On many of the larger islands, there are year-round communities of lobstermen and fishermen as well as a number of vacation and retirement homes such as the one profiled in our case study. All share a common heritage of dependence on kerosene lamps, hand-pumped water, and private diesel generators. All this is changing now, thanks to a growing acceptance of stand-alone, battery-equipped PV systems by summer and winter residents alike.

It was on one of these small islands in Frenchman's Bay that television newsman Jack Perkins and his wife decided to build a retirement home. The home features 1,800 square feet of living area and includes almost every creature comfort you might want. With cooking, refrigeration, and space heating supplied by propane, the loads on the home's photovoltaic system were calculated to be about 2.5 kilowatt-hours per day or about 75 kilowatt-hours per month. This includes power for the deep-well pump, a dishwasher, a washing machine, a gas-fired clothes dryer, lighting, and miscellaneous appliances, as well as communications.

Consulting a source of insolation data, we find that the annual average daily peak sun hours at this site is 3.5 kilowatt-hours per square meter falling on a surface inclined at 45 degrees and facing true solar south. We also note that the daily insolation for the "winter" months of November through March averages about 2.7 kWh/m^2 per day with a low of 2.23 kWh/m^2 per day in December. Insolation values for the site at 44.2° north latitude expressed in kilowatt-hours per day (peak sun hours) are shown in table 8-1.

Table 8-1. Insolation Values by Month and Season for Bar Harbor, Maine

Season	Month	Average Insolation per Month (kWh/m^2/day)	Average Insolation per Season (kWh/m^2/day)
Winter	Nov.	2.32	2.69
	Dec.	2.23	
	Jan.	2.52	
	Feb.	2.95	
	March	3.45	
Spring	April	4.00	4.20
	May	4.40	
Summer	June	4.60	4.51
	July	4.56	
	Aug.	4.39	
Fall	Sept.	3.87	3.64
	Oct.	3.40	
	Year Average	3.56	

The system designer, David Sleeper of Falmouth, Maine, reviewed the load calculations and insolation data with the owners. As a result of this discussion, the owners decided not to size the system for the worst-case month of December but to rely on backup power from a 7.5-kilowatt propane-fired generator. This additional power would supplement the daily photovoltaic harvest as required, especially during the winter "doldrums." The owners reasoned that, while the house was designed for year-round living, its first years of use would be primarily during the spring, summer, and fall. If and when they should decide to winter there, they could always increase their system's array and battery capacity. Knowing that the exposure to the North Altantic that makes for unbelievable summer sunrises on this island site can and does also deliver some rather rough winter nor'easters, Dave Sleeper was quick to agree.

Once the decision was made to rely (at least initially) on supplemental power from the generator in the winter months, Dave decided, based on his knowledge of insolation at the site, to use a figure of 3 peak sun hours per day to determine the desired array output. Dividing the estimated 2,500-watt-hour-per-day load by this figure, we see that a system capable of producing 833 peak watts was needed (2,500 ÷ 3 = 833.33 W_p).

Array Sizing Using the Watt-Hour Method

The array at the Maine Island house consists of 32 Mobil Solar Ra-30 modules, each with a rated output of 30 peak watts at 15.5 VDC and 1.94 amps (at STC of 1,000 W/m^2 and 25°C). When the correction in module output is made for temperature (as must be done if the sizing is done in watt-hours), we arrive at an output voltage of 13.87 VDC and a nominal (peak) wattage output of 27.48 watts (at 1,000 W/m^2) under a NOCT of 46°C (115°F). The modules are arranged in an array of eight parallel strings of four series-connected modules each, for an easily divisible total of 32. The array output should thus be 879 peak watts at an operating voltage of approximately 55 VDC under average site conditions.

Reviewing the insolation data again, we find that this 879-W_p array will deliver an average of 3.97 kWh/day during the summer months of June through August, 3.2 kWh/day during the months of September and October, and 3.69 kWh/day in April and May. Overall, the (nonwinter) average daily output is approximately 3.6 kilowatt-hours.

The chosen design value of 3 hours of peak sun per average day appears somewhat conservative for the nonwinter season when compared to these figures. However, this is good design practice, since electrical demand seldom shrinks and usually grows. Owners' use patterns change, relatives and guests visit, seasons are extended, and so on. In addition, it is also difficult to assign exact numbers to the internal losses in the system (due mostly to inverter operation and battery round-trip efficiency) because the amount and the timing of the owners' use of each individual load cannot be precisely predicted.

Let's look a little closer at the numbers and identify these internal losses to see just how much they affect array-sizing requirements. In chapter 3, I said that the amp-hour method of array sizing is preferred for SA photovoltaic systems that charge a battery if they are under 48 volts. Since this system is to operate at 48 volts, we'll do the calculations in both watt-hours and amp-hours to demonstrate the two procedures and compare the results, starting first with the watt-hour method.

Let's assume that we can take the average net daily load demand, which was calculated to be 2.5 kilowatt-hours or 2,500 watt-hours, and determine that 40 percent of this load (1,000 Wh) will occur during sunlight hours and be satisfied directly from the array, and that the remaining 60 percent (1,500 Wh) will be satisfied from storage. Let's further assume that 60 percent of this average daily load (1,500 Wh) will require inverter operation while the remaining 40 percent (1,000 Wh) will be handled by direct current. From this data we are able to make calculations and arrive at the net average daily system loads as shown here.

DC loads served direct from array	400 Wh
DC loads served from storage	600 Wh
AC loads from array through inverter	600 Wh
AC loads from storage through inverter	900 Wh
Total	2,500 Wh

To these figures, we must assign the internal system losses appropriate to the watt-hour sizing method. For DC loads served directly from the array, we add 5 percent for regulation and wire losses. For DC loads served from storage we add 25 percent, which includes the 5 percent for regulation and wire losses and 20 percent for battery round-trip efficiency. For AC loads served directly from the array we add 20 percent, which includes the 5 percent for regulation and wire losses and 15 percent for the *average* efficiency of a good-quality inverter. And for AC loads served from storage we add 40 percent, which includes the 5 percent for regulation and wire losses, the 20 percent for battery efficiency, and the 15 percent for average inverter efficiency. Taking these losses into account, we then arrive at the actual load requirements as shown.

DC loads served direct from array	400 Wh × 1.05 =	420 Wh
DC loads served from storage	600 Wh × 1.25 =	750 Wh
AC loads from array through inverter	600 Wh × 1.20 =	720 Wh
AC loads from storage through inverter	900 Wh × 1.40 =	1,260 Wh
	Total	3,150 Wh

The total daily load shown is 3.15 kilowatt-hours. This compares very well with the average array output of 3.6 kWh/day projected initially for the period of April through October. It includes enough additional array output (approximately 10 percent) during spring and summer to assure recharging of storage while simultaneously satisfying system loads after a

Table 8-2. Array Output by Month and Season

Season	Month	Average Array Output per Month (kWh/day)	Average Array Output per Season (kWh/day)
Winter	Nov.	2.04	2.37
	Dec.	1.96	
	Jan.	2.21	
	Feb.	2.59	
	March	3.03	
Spring	April	3.52	3.69
	May	3.87	
Summer	June	4.04	3.97
	July	4.01	
	Aug.	3.86	
Fall	Sept.	3.40	3.42
	Oct.	3.00	
	Year Average	3.13	

prolonged period of no-sun days, when the battery bank has reached its maximum depth of discharge. The (average) daily output of the system's 879-peak-watt array in watt-hours is shown in table 8-2.

During the fall, some additional power input will be required from the backup generator, since October's average daily array output is only 3 kilowatt-hours and the average daily load requirement is 3.15 kilowatt-hours. However, the system's backup generator should, in any case, be brought on-line for a minimum of 1 to 2 hours every month in keeping with the manufacturer's recommendations for exercising the machine. When this additional generator power is added to the output from the array, there will be a modest power surplus available even during October to help accommodate all those uninvited guests.

Array Sizing Using the Amp-Hour Method

If we perform these calculations once again using the amp-hour method of array sizing rather than the watt-hour method, what will the numbers look like? Let's start with the loads first. With a nominal system voltage of 48 VDC specified, the system's net load calculated at 2.5 kWh/day amounts to 52 amp-hours. The various categories break down as shown here.

DC loads served direct from array	8.33 Ah
DC loads served from storage	12.50 Ah
AC loads from array through inverter	12.50 Ah
AC loads from storage through inverter	18.75 Ah
Total	52.08 Ah

Now we must include the appropriate internal system losses, which are the same as in the previous sizing calculation done in watt-hours except for the battery bank round-trip efficiency. As you are by now aware, the battery bank is considered 100 percent efficient in terms of amp-hours in versus amp-hours out when the amp-hour sizing method is used. The breakdown of the relevant system losses is shown below. As you can see, the actual daily total load is 59.38 amp-hours.

DC loads served direct from array	8.33 Ah × 1.05 =	8.75 Ah
DC loads served from storage	12.50 Ah × 1.05 =	13.13 Ah
AC loads from array through inverter	12.50 Ah × 1.20 =	15.00 Ah
AC loads from storage through inverter	18.75 Ah × 1.20 =	22.50 Ah
	Total	59.38 Ah

We know from the sizing examples presented in chapter 3 that the Mobil Solar Ra-30 module will deliver a current of 2 amps under full sun. The array has eight parallel strings of four Ra-30 modules each and will deliver 16 amps (at 48 VDC nominal) or 16 amp-hours per hour of full sun. The array output in amp-hours is presented in table 8-3.

The amp-hour sizing method also indicates a shortfall of PV-generated power in October. However, when the backup generator's power input that results from the manufacturer's recommended exercising schedule is added to the array output above, we find that we again have a modest surplus of power in the spring and fall.

You will notice that the two sizing methods don't produce exactly the same results. At the level of detail with which we have performed these calculations, the watt-hour method will deliver an array output that is slightly more optimistic than the amp-hour method. This is because the watt-hour method of sizing treats the array as operating at its maximum power point, which is its optimum (most efficient) voltage. This is somewhat higher than what will actually happen in the field (in systems without max-power tracking), because during periods of low state of charge the batteries will pull the array voltage down and away from the max-power point. This will bring about an average array voltage that is somewhat lower than the peak-power-point voltage.

The amp-hour method, on the other hand, treats the array as always operating at the nominal voltage rating of the system. This is in prac-

Table 8-3. Array Output by Month and Season for the Maine Island House

Season	Month	Average Array Output per Month (Ah/day)	Average Array Output per Season (Ah/day)
Winter	Nov.	37.12	43.10
	Dec.	35.68	
	Jan.	40.32	
	Feb.	47.20	
	March	55.20	
Spring	April	64.00	67.20
	May	70.40	
Summer	June	73.60	72.27
	July	72.96	
	Aug.	70.24	
Fall	Sept.	61.92	58.24
	Oct.	54.56	
	Year Average	56.85	

tice somewhat low, because the array frequently operates above the nominal voltage rating. If the system is designed properly so that array output and storage capacity are matched to the load demands, the average operating voltage of the array will often be somewhat above this value.

What the array will actually deliver in the field is likely to fall slightly below the output projected using the watt-hour method and above the output projected using the amp-hour method. However, the results are close enough to each other that in the majority of cases the use of either method will produce satisfactory results. I recommend the use of the amp-hour method in this case because it is less complicated, and any sizing error will most likely result in a slightly larger array than necessary, which is desirable in this type of system.

Array-sizing calculations could be performed to a greater level of detail than we have demonstrated here. However, I don't feel this is necessary for basic home power systems. It may even be a waste of time, because there are some system parameters that cannot be more carefully defined. For example, the *real* load placed on a system and the *actual* use patterns followed by people after they start to live in a solar electric house simply cannot be precisely determined in advance. The best you can hope for in an initial load estimation is to be within 5 or 10 percent of the actual load. That level of estimating accuracy is very good even when it comes to designing your own system rather than a system intended to translate someone else's lifestyle into kilowatt-hours per day.

Battery Storage Sizing

Six full days of battery storage at an MDOD of 25 percent and a DDOD of 3 to 5 percent was settled upon as the design objective for the battery bank in this house. The system's designer specified 48 Delco 2000 shallow-cycle photovoltaic batteries, each with a rated capacity of 105 amp-hours at 12 volts. The batteries were arranged in 12 parallel strings of four series-connected batteries to produce 48 VDC.

Now let's see how this storage size was determined. At 2.5 kilowatt-hours per day, six days' full load would be 15 kilowatt-hours. At 48 volts, this translates into a load of 312.5 amp-hours for the design period. If we are not to discharge below 25 percent after six days, the rated battery capacity must be 1,250 amp-hours (312.5 ÷ 0.25).

Each Delco 2000 has a rated total capacity of 105 amp-hours. When we divide the required total capacity by the individual battery capacity (1,250 ÷ 105), we find that we need 11.9 batteries. Since we cannot use a fraction of a battery, we round the result up to 12 batteries in parallel.

Now, for good measure we check the DDOD of our battery bank. Knowing that we have a daily system requirement of 52 amp-hours and a total battery capacity of 1,250 amp-hours, we can quickly calculate the DDOD (52 Ah ÷ 1250 Ah = 0.0416). We see that in satisfying our estimated system requirement the battery bank will experience a DDOD of only 4 percent, which is well within the manufacturer's recommended levels.

The system's operating voltage was set at 48 VDC. The battery bank is configured in 12 parallel strings of four batteries each to deliver a nominal input voltage to the inverter of 48 VDC, with an absolute storage capacity of 1,250 amp-hours and a usable storage capacity of 312.5 amp-hours (15 kWh) at 25 percent MDOD. This translates almost exactly into six days' storage under the design load of 2.5 kilowatt-hours per day. Full dependence on the batteries for longer than six days would result in an MDOD of more than 25 percent, which is not recommended if the batteries are to last their full rated lifetime.

The designer used the shallow-cycle batteries for this application because the units are physically smaller and lighter than their deep-cycle counterparts. Size and weight were very important here, since the residence was built on an island site and everything had to be moved in first by boat and then by hand. When designed and operated properly, the shallow-cycle battery bank will deliver satisfactory service. However, a system built around a deep-cycle unit will likely offer cost advantages.

If the battery storage level in the Maine Island house ever falls below 75 percent of capacity (25 percent DOD), an automatic transfer

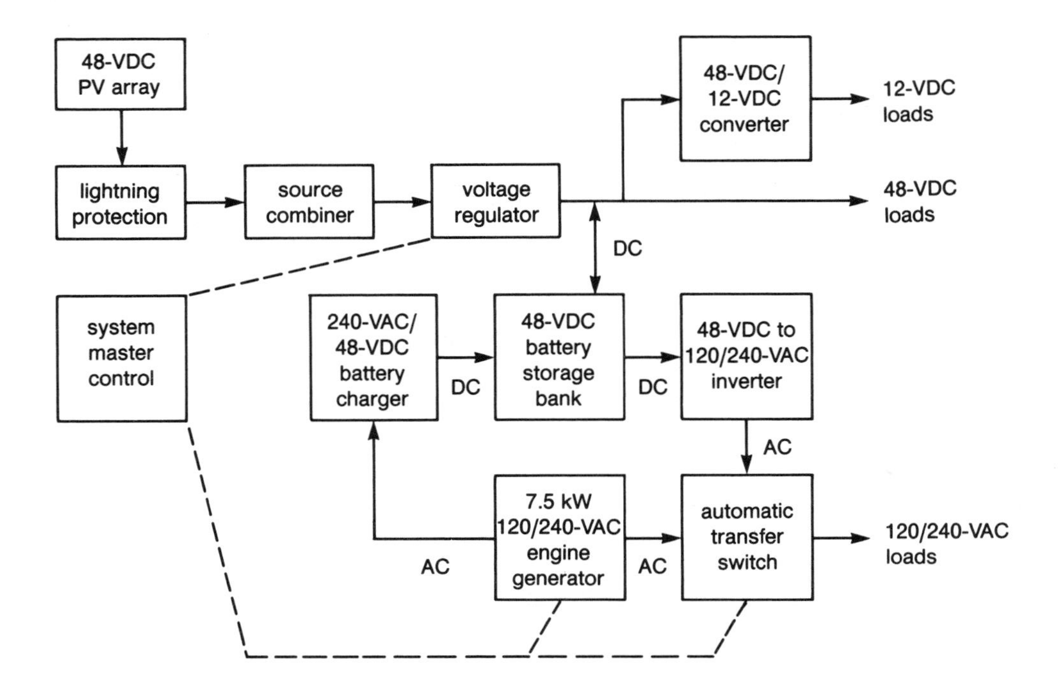

FIGURE 8-3

A block diagram of the stand-alone PV system at the Maine Island house.

switch will activate the 7.5-kilowatt propane-fired generator, which transfers power to the batteries via a 150-amp battery charger. (For a temporarily heavy load, the switch can be manually overridden.) In practice, however, the generator is seldom used except to keep it exercised.

Alternating current electrical service in the Maine Island house is provided by a Best 48-VDC input, 5-kilowatt inverter with 120/240 VAC output. The inverter features automatic on-demand starting. The DC loads in the house include 48-volt incandescent and fluorescent lights, both of which run directly off the battery bank, and a 12-VDC Marine VHF radio and a high-frequency-link private telephone system, both of which depend upon a small DC-to-DC converter to step down the 48-VDC output from the battery storage bank.

The Maine Island house is a model photovoltaic installation that has proven to be extremely satisfactory to its owners in a situation where dependence on the central uitlity grid is not possible. It typifies the pioneering work done in the part of the United States that is touched first by the rays of the morning sun. On Monhegan Island, which is 12 miles due east of the coastal town of Booth Bay Harbor, there are now over two dozen PV systems, including one to power the island's post office. These small Maine islands such as Monhegan really comprise the first PV-powered communities in the United States. For this reason they were chosen to be represented at Walt Disney World's EPCOT Center in its permanent alternative energy exhibit.

The Hudson Valley House

Early in 1982, my firm was commissioned by a private client to design a full-size, year-round SA photovoltaic home in New York's Hudson River Valley. This was to be a luxury home—the client was interested in conservation, but not in austerity or self-denial. He also preferred to be completely independent of the utility grid. Although the chosen site was rural, a utility connection would have been entirely feasible had it been desired. Instead, the steady supply of electricity to loads was designed to be entirely the joint responsibility of the PV array, the battery bank, and a propane-fired backup generator.

The house incorporates approximately 2,400 square feet of floor space and includes a living room, dining room, kitchen, library, workshop, study, master bedroom, two guest bedrooms, 2½ baths, pantry/laundry area, storage areas, and utility space. A two-car garage with an upstairs loft is connected to the house by a passageway and the room housing the PV system equipment and the battery bank. Despite this ample amount of space, heating loads are light, thanks to earth sheltering, heavy insulation (R-40 for exposed walls, R-60 for the roof), and passive gain through south-facing windows and sliding glass doors. Passive design also eliminates the need for air conditioning. Both space heating and domestic water heating are active solar, with propane backup.

In the Hudson Valley house, propane is also used for clothes drying, cooking, and refrigeration. Since this was a new house, and because the owners are committed to conservation and a substantial degree of photovoltaic independence, all new appliances were chosen with energy efficiency the foremost criterion. Load management, however, remains entirely voluntary. Despite the spaciousness of this house and its many electrical amenities, the average net daily load demand on the PV system was kept to a conservative 10.75 kilowatt-hour-per-day total. We'll look at the design and sizing of the PV system for this house in enough detail to describe the approach utilized, but we will not repeat each phase of the calculations to the depth of detail presented previously.

The Hudson Valley house, like the Maine Island house, was designed for a retired couple, and the daily load profiles in terms of time of use are likely to be quite similar in the two houses. Using the same load distribution and loss parameters presented in the case study of the PV house in Maine, let's see how the numbers break down.

The average net daily load demand was calculated to be 10.75 kilowatt-hours. It was determined that 40 percent of this load will occur during sunlight hours and be satisfied directly from the array and that the remaining 60 percent will be satisfied from storage. It was further determined that 60 percent of this average daily load will require inverter operation and the remaining 40 percent will operate on unmodified direct current. From these basic figures, we calculate the net average daily system loads for the Hudson Valley House to be as shown in the calculation on the following page.

DC loads served direct from array	1,720 Wh
DC loads served from storage	2,580 Wh
AC loads from array through inverter	2,580 Wh
AC loads from storage through inverter	3,870 Wh
Total	10,750 Wh

As in the previous case study, we must also calculate internal system losses appropriate to the watt-hour sizing method and add them to the load. For DC loads served directly from the array, we add 5 percent for regulation and wire losses. For DC loads served from storage we add 25 percent, which includes the 5 percent for regulation and wire losses and 20 percent for battery round-trip efficiency. For AC loads served directly from the array we add 20 percent, which includes the 5 percent for regulation and wire losses and 15 percent for the *average* efficiency of a good-quality inverter. And for AC loads served from storage we add 40 percent, which includes the 5 percent for regulation and wire losses, the 20 percent for battery efficiency, and the 15 percent (average) inverter efficiency. Taking these losses into account, we then arrive at the actual load requirements shown below. As you can see, the total daily load including system losses is 13.54 kilowatt-hours.

DC loads served direct from array	1,720 Wh × 1.05 =	1,806 Wh
DC loads served from storage	2,580 Wh × 1.25 =	3,225 Wh
AC loads from array through inverter	2,580 Wh × 1.20 =	3,096 Wh
AC loads from storage through inverter	3,870 Wh × 1.40 =	5,418 Wh
	Total	13,545 Wh

The photovoltaic array we designed for the Hudson Valley house utilizes Mobil Solar Energy Corporation's 4-foot by 6-foot Ra-180 PV modules. Each module is made up of 432 ribbon cells and has a rated output of 180 peak watts. The PV system will have a propane-fired generator to fall back on during the winter "doldrums." However, we still based the array sizing on the average daily winter insolation, since this house will be occupied full time and the clients wish to minimize generator input to the system.

Consulting a resource on insolation data, we find that the average winter peak sun hour figure for the site is 3.3. Dividing a 13,545-watt daily load by 3.3, we find that the required array output is just over 4,104 peak watts, or 4.1 peak kilowatts. In this system, we are using large-area, high-wattage PV modules, each of which has a peak output of 180 watts.

PHOTO 8-5

These large-area, roof-integrated PV modules from Mobil Solar are installed directly over the roof rafters at the Hudson Valley house; the completed array is the finished roof.

Therefore, we determine the number of modules needed through the following equation:

$$\frac{4{,}104\ W_p}{180\ W_p/\text{module}} = 22.8\ \text{modules}$$

Since we cannot work with a fraction of a module, we round up to 23. However, 23 is a prime number, which means that it cannot be evenly divided electrically into array strings and cannot be symmetrically configured on the roof in anything but one row of 23 modules. Fortunately, adding just one more module gives us 24. As we'll see in chapter 9, this creates an ideal array in terms of electrical configuration and physical mounting options.

Thus, the array was specified to contain 24 Mobil Solar Ra-180 modules. The system voltage was specified at 120 VDC because of the size of the array, the battery bank, and the current levels that would be present in the system. To deliver 120 VDC, 12-volt (nominal) Ra-180 photovoltaic modules were selected. Consulting the manufacturer's literature at that time, we found that after correction for temperature, eight series-connected modules would be required per string to produce this voltage. The array was electrically configured in three parallel strings of eight modules each and integrally mounted on the south-facing roof at an angle of 50 degrees, somewhat in excess of site latitude. This angle was chosen to favor the winter sun, since winter is the time of lowest available insolation.

Battery storage was the next step of the design process. Again, the 12-volt Delco 2000 battery, with its 105 amp-hours of storage, was chosen for the job. For two full days of battery dependence, at the recommended 15 percent depth of discharge, we selected 140 batteries. The batteries are paralleled in 14 series strings of ten series batteries each.

In order to determine the DDOD of the batteries in the battery bank, we first calculated the total rated capacity of the batteries in terms of kilowatt-hours. Multiplying the nominal voltage by the rated amp-hour capacity (12 × 105) we found that each battery is able to store 1,260 watt-hours (1.26 kilowatt-hours).

Since the manufacturer's recommended DDOD for this battery is 15 percent, we multiplied the total rated capacity by 0.15 and found that each battery had a total usable capacity of 0.189 kWh. Multiplying this number by 140 (the number of batteries in the battery bank) we concluded that the TUC of the battery bank as a whole would be 26.46 kilowatt-hours, a fraction less than the 27.08 kilowatt-hours that represents two days' estimated load for the house. This slight difference can be covered by a minor effort at voluntary load management.

Since the estimated daily load is 13.54 kilowatt-hours, five days of complete dependence on the batteries would mean drawing 67.7 kilowatt-hours out of the total 176.4 kilowatt-hours of storage in the battery bank. This would take the batteries down to almost a 40 percent DOD (67.7 ÷ 176.4 = 0.38), well below the manufacturer's recommended level.

PHOTO 8-6

The battery bank at the Hudson Valley house features 140 Delco 2000 batteries and two 150-amp battery chargers (at left).

In fact, the batteries in the Hudson Valley house (like those in the Maine Island house) are protected from too drastic a depletion of charge by an automatic generator demand start switch. In this case, the switch activates a 12.5-kilowatt Kohler generator, which feeds battery storage through two 150-amp Lester battery chargers.

Actually, there are two control modes in which the generator can be automatically turned on. Both are "field-settable," that is, they can be selected, adjusted, and reset on-site at the owner's discretion. One option causes the generator to come on when the battery DOD reaches a certain level (for example, 15 percent), but only during a "time window" determined by a timer built into the system. This could be set to happen, for instance, between 4:00 and 6:00 P.M., during a heavy load period. The other option calls for generator activation upon arrival at a certain DOD regardless of the hour, in which case the timer is overridden. Both modes have a manual override that allows the generator to be started at any time the owners desire (such as for exercising).

Overcharging of the batteries from the array is prevented through the use of a voltage regulator. Should any gassing occur due to a slight overcharge, hydrogen is vented to the outdoors through a specially designed manifold.

One of the noteworthy features of the battery-to-load service in the Hudson Valley house is the coexistence of seperate AC and DC distribution systems. All lighting, as well as the pumps and controls for the radiant floor space-heating system are served by a 120-VDC line. The remainder of the loads draw power from the 120/240-VAC inverter output. The reason for the DC/AC division was so that direct current could be used where alternating current is not absolutely necessary, reducing the burden on the inverters and minimizing inversion losses. The DC line voltage could of course have been lower, but given the wattage demand of the DC loads, such a decision would have mandated heavier and more expensive wiring. Hence the choice of 120-VDC service, as provided by the ten-series battery arrangement.

At the time the Hudson Valley house was designed, power inversion represented a special set of challenges. Because the house was to have an SA photovoltaic system, there was particular awareness of the need for an ample margin of safety in this potentially most vulnerable part of the entire PV system. A single, oversized inverter would have been an inefficient approach to the problem. Our solution was to divide the job between a pair of Best 6-kilowatt, self-commutated inverters. Each inverter handles roughly half of the major AC loads. If one of the units should fail, the other is capable of taking over all requirements provided a measure of self-imposed load management is applied to load demand.

There still remained the thorny problem of those small loads that require a 24-hour power supply but that could not justifiably be served by the main inverters. As we discussed before, the lighter the loads on an inverter, the less efficient it becomes. The "tare loss" of power required just to run the inverter while it idles or serves a minuscule AC load represents a

PHOTO 8-7

The balance-of-systems equipment for the Hudson Valley house was assembled and wired on a large panel board in the shop and shipped as a complete unit to be installed at the job site.

waste of valuable stored photovoltaic electricity. It's far more efficient to allow an inverter (or, in this case, a pair of inverters) to remain in an "off" mode until activated by a demand signal from one of the larger loads. But what to do about those marginal loads associated with essential, round-the-clock devices such as the solar heating system controls and pumps, pilotless gas appliance ignition, an overhead garage door remote signal receiver, and the security system?

The solution to the problem in the Hudson Valley house was the installation of a separate "dedicated inverter" circuit, through which a 250-watt inverter (one-twenty-fourth the size of either of the two main units) provides constant power for these and other small loads on a 24-hour basis. With such a low-output inverter, idling losses are minimal and easily acceptable. These are the sort of design details that can make stand-alone photovoltaics viable in a large house without overdesign or sacrificing of amenities.

Recent advances in stand-alone inverter technology since the house was completed have made the need for a dedicated inverter less critical. For example, Heart Interface now manufactures 2.5-kilowatt SA inverters that are cascadable. Up to four of these units can be connected to the same master control board to provide up to 10 kilowatts of load capability with 40 kilowatts of total (gross) surge capacity. Each unit draws only 1.8 watts in the idling mode. We can expect to see even more impressive innovation in stand-alone inverters in the near future.

Because of the advances made in photovoltaic system hardware in the years since the Hudson Valley house was built, if I were asked to design the house's power system today I would do some things differently. For example, in the larger SA residences currently being designed by my

PHOTO 8-8
The south face of the Hudson Valley house.

firm, I am specifying standard AC distribution from the inverter to all loads. Considering the recent improvements in SA inverter efficiency, the problems and additional cost involved with separate DC power distribution and special DC loads now outweigh its benefits in systems of this size.

If I were designing the system again I would also use deep-cycle batteries. Although the shallow-cycle units have performed quite well, deep-cycle batteries have advanced over the years. Their use can represent a cost savings in installation because they provide more kilowatt-hour storage capacity per single unit. This means fewer to handle, provide racks for, install, and wire. In addition, the deep-cycle units now on the market can, if properly selected, provide a significant advantage in life-cycle costing (as can be seen from the figures found in table 4-3). The initial capital investment may be higher but the number of sets of batteries needed over a 20-year period will be less, which means long-term savings on both components and replacement labor.

The Hudson Valley house is not typical of today's SA photovoltaic residences. Most are quite a bit smaller, and many are retrofit installations used seasonally. It was, to my knowledge, the first time a year-round home of its size was designed from the ground up to be fully powered by a stand-alone PV system.

The industry has advanced significantly since the Hudson Valley house was completed. Costs have come down, efficiencies have improved, new products have been introduced, and control electronics have become more sophisticated. Many options are available as off-the-shelf items today, whereas most of the controls in the Hudson Valley house had to be custom-built. Despite the pioneering nature of the project, all of the systems have functioned well over the years and the owners have *never* been without power. That is a fact that pleases me just about as much as it pleases them.

Taken together, the diverse SA systems described in this chapter vividly illustrate the broad range of applications for this technology. Stand-alone photovoltaics can form the basis of a small, relatively simple power system such as that enjoyed by Bart and Charlene Higgins in their log cabin located in the remote Labradoran fishing village of Paradise River. They can also be used to provide energy independence for residents of a spacious, luxury home equipped with the latest in modern conveniences, such as the Hudson Valley house. In short, stand-alone PV systems can be designed to answer just about any residential power need.

CHAPTER NINE

INSTALLING YOUR PHOTOVOLTAIC SYSTEM

In larger photovoltaic systems with DC voltages greater than 48 volts, the choice of a PV module determines string length. String length, in turn, strongly influences both the electrical characteristics and the physical size and shape of the array. Thus, module selection immediately establishes the number of options available to the architect and system designer when sizing and laying out the configuration of the PV array. This is especially critical for arrays that will be integrated into the structure of a building.

The importance of module voltage and string length increases with the size of the array and its operating voltage. For example, a system that requires 13 modules in a series string to achieve the necessary operating voltage can be symmetrically configured in only one way: rows or columns of 13 modules. A 14-module string can be arranged in two ways: 1 times 14 or 2 times 7. A 12-module string can be configured in many ways—1 times 12, 2 times 6, 3 times 4, L-shape, and so on—to allow the array to be physically configured to fit the building and not the other way around.

It is easy to see from this that the more ways a string can be evenly divided, the more different array configurations are possible, and that array strings using prime numbers of modules should be avoided. Generally, the higher the module operating voltage, the fewer modules will be required to make up a given string and the more flexible the array design can be in terms of both physical configuration and power output options.

When the need to integrate a photovoltaic array is added to the many, often conflicting, issues already involved in the house design process, this kind of flexibility becomes critical to the efforts by the architect or designer to create aesthetically pleasing living environments that people will enjoy.

Our experience in designing and constructing photovoltaic systems has led us to conclude that high-density, high-efficiency arrays using large-area (at least 2-square-meter), high-voltage PV modules (24 to 48 VDC nominal or greater) will eventually become the industry standard for residential and many commercial applications. Mobil Solar currently offers such a large-area module, the Mobil Ra-180. Other manufacturers have recently begun to combine many of their small modules as unframed laminates into larger-area, higher-voltage panels.

The design of large-area, high-efficiency modules is an important priority because the support structure for a photovoltaic array is expensive. This is especially true if the supporting structure is a house or other building. It is very hard to justify building a bigger house just to support a large, low-efficiency PV array.

The goal, then, is to select photovoltaic modules that provide flexibility in array sizing and configuraton while delivering the maximum power that is economically practical from a given array area. This keeps construction costs low and reduces the architectural impact of the array, allowing the designer more freedom and making more roof aperture available for other solar applications.

Array Mounting

Photovoltaic array installations fall into two general categories: roof mounted and ground mounted. In addition, a distinction may be made between arrays installed as a part of new construction and those retrofitted on an existing structure. At present, most retrofit installations have been limited to relatively low-output arrays that address modest load requirements such as those found at a remote vacation home or on a farm or ranch where water pumping, basic lighting, and communications may be the sole electrical loads.

PHOTO 9-1

There are now many thousands of cabins powered by small ground-mounted PV arrays.

Four generic methods have been developed for the mounting of photovoltaic arrays: *rack mounting*, where the modules are mounted on a support structure on the ground or above the flat roof of a building in a plane different from that of the roof; *stand-off mounting*, where the modules are mounted on a frame that stands 2 to 6 inches above the finished roof of a building, in the same plane as the roof; *direct mounting*, where the modules are mounted directly on the plywood roof sheathing; and *integral mounting*, where the PV modules are mounted directly on the building's structural roof members, completely displacing the conventional roofing system.

Ground Mounting

Unless your home lacks sufficient south-facing roof area, you will most likely want to consider a roof-mounted array for your residential photovoltaic system. When compared to a well-designed roof-mounted installation, a ground-mounted array is less elegant, requires longer wire runs, is more likely to be shaded, and occupies land that could be used in other ways. Fencing is desirable to keep out animals, children, and trespassers. Wiring must be protected from damage by rodents. Also, weeds and drifted snow may need to be cleared from time to time, though falling snow will usually slide off if adequate ground clearance is provided.

However, there are some advantages to ground mounting, particularly if adjacent open land is available at little or no cost. A ground-level photovoltaic array is easily accessible during installation, as well as for cleaning and occasional maintenance after system start-up, and may easily be expanded to incorporate additional modules.

Ground-mounted structures usually are built with frame members of steel, aluminum pipe, angle, or (sometimes) wood, very similar to those used in the rack approach described later. Whereas the roof-mounted rack is secured to the building, the ground-mounted support structure is usually secured to a foundation of concrete piers, although some people have used railroad ties or precast concrete such as parking lot curb bumpers as an array base. As with the roof-mounted support rack, the ability to stand up to predicted wind loads is a prime consideration with any ground-mounted support structure.

Tracking the sun can increase your array output by as much as 40 percent in areas with many hours of daylight and high direct-beam insolation. However, this is not free additional energy. The increased cost of the tracking system must be factored into the energy equation along with any future maintenance requirements. Whether a tracking ground-mounted support is the right choice for your application can be decided only by a careful consideration of all the factors involved. As PV module costs continue to come down, the added cost of tracking a residential PV power system will become increasingly harder to justify.

If tracking your residential photovoltaic array appeals to you, the simplest, most trouble-free, and least expensive of these tracking concepts is the passive freon tracker pioneered by Steve Baer at Zomeworks. This approach uses two freon canisters. When one is exposed to the sun, the freon boils off and is driven inside the other one, where it condenses again.

This action creates a weight imbalance that serves to move the PV panels attached to the tracker along the sun's path over the course of the day. Passive, freon-driven, pedestal-mounted trackers are available from Zomeworks for small arrays of up to eight or ten modules.

Most ground-mounted photovoltaic arrays that supply power to buildings are either quite small (remote cabins or camps) or extensive (for example, Beverly, Massachusetts, High School and Natural Bridges National Monument in Utah). The vast majority of residential PV applications will lie between these extremes in size and, in most cases, will be best served by a roof-mounted array. Fortunately, the roof area of a typical house is generally sufficient to accommodate an array capable of handling most residential loads.

The Rack Mount

As we learned in chapter 3, to obtain the highest annual array output, the angle at which the modules face the sun should roughly equal the latitude of the site. Chicago, for instance, is at about 42° north latitude; thus, a photovoltaic array in the Chicago area would be ideally angled at approximately 42 degrees from the horizontal. In most parts of the country, designers of solar water heating systems and the occasional PV retrofit have been fortunate to find that roofs are often pitched at an angle closely corresponding to the site latitude.

Of course, a house expressly designed for photovoltaics is given an optimum roof pitch before it is off the drawing board. Flat and gently pitched roofs are, however, not uncommon, and in some areas of the Southwest they are often the rule. Rooftops that are flat or inadequately pitched for photovoltaics will require rack mounts with a structure very similar to that used for ground mounts.

In a rack mount, the modules are supported at the desired altitude angle by a triangulated understructure commonly built of aluminum, steel angle, or steel pipe. Sometimes an easily assembled and adjusted system of manufactured pipe fittings is used, such as the Speed Rail and Kee Klamp systems favored by many installers of solar thermal collectors. Occasionally, wooden framing is used. The rack-mounted array stands free above the roof, replacing none of the traditional roofing materials. The frame members are usually lag-bolted to the rafters beneath the roof sheathing or through-bolted into solid blocking placed between the rafters. Waterproofing these penetrations in the roof membrane is an important consideration.

When rack mounting is necessary, it will involve some additional expense as well as increase the weight and design loads of the entire installation. The additional weight is unlikely in and of itself to pose any danger to a sound roof if the support structure is properly attached. However, the elevation of modules above the roof will require that maximum anticipated wind velocities be carefully taken into consideration during the design of the array. A simple wind or snow load (downloading) is one phenomenon to be considered. Often more important is the negative load or uploading

PHOTO 9-2

The PV array at the Santa Fe house was easily rack-mounted over the flat roof.

created by the lift effect that can result when the wind blows from the rear against a sloped rack-mounted array.

There are some decided advantages to the rack mount. We know that photovoltaic cells undergo some loss of efficiency when operated at high temperatures, and the free-air ventilation of both front and rear module surfaces afforded by the rack mount helps to keep the modules cooler. Also, the exposed rear surfaces in a rack mount offer ease of access to electrical connections between modules during installation and for testing and maintenance.

Another rack-mount benefit, which perhaps only the more dedicated and technically demanding photovoltaic users might wish to pursue, is the ability to adjust the array angle to account for seasonal differences in the direction of solar radiation. In the winter, an increase of 15 degrees in the altitude angle will improve the seasonal efficiency by approximately 5 percent. In the summer, the same improvement may be achieved by subtracting 15 degrees and bringing the array closer to a horizontal position. However, most PV users are content with the knowledge that, during the summer, the marginal efficiency drop associated with imprecise angling is compensated for by the greater abundance of sunlight.

Many photovoltaic manufacturers and distributors have preengineered rack-mounting systems available for their modules. These kits are most often made of aluminum extrusions and are shipped with all of the pieces precut and predrilled, ready to assemble. There are several advantages to buying a preengineered rack. Besides being prefabricated and ready to install, it is designed expressly for the mounting of PV modules and built to withstand the forces of wind and snow loading imposed on rack-mounted arrays.

ARCO Solar and Mobil Solar both offer preengineered racks for their modules, and Photron has designed a prefabricated rack-mounting system that can accommodate PV modules from several manufacturers.

Most of the preengineered rack-mounting systems available allow the altitude angle of the array to be varied, providing flexibility during the initial installation and allowing those who wish to adjust their array seasonally the ability to do so.

Another rather nice preengineered option for rack mounting is the family of support hardware produced by Leveleg. Originally designed for solar thermal collectors, the system replaces the frame, the support, and the mounting feet of a traditional rack mount with telescoping supports that are adjustable to allow field setting of the altitude angle. The top of the support fastens to the module's frame rail and the bottom is lag-bolted into the roof. Leveleg has recently added a theft-resistant hardware option that should deter thieves in remote areas. The advantages of the system are best obtained in small to medium-sized home power systems installed in areas that do not receive much snow.

Since photovoltaic modules are far less tolerant of shading than solar thermal collectors, the angle of sunlight *at each season of the year* must be taken into consideration when a ground or rooftop rack-mounted array is designed. Shadowing can result not only from the interference of trees, buildings, or other obstructions but also from positioning one rack of modules too close to another.

In the winter, the sun's rays fall more obliquely, and shadows are much longer. Partial shadowing of a module in which cells are connected in series will fully incapacitate the module for the duration of the shadow. For series-connected modules the effect will be the same, with the entire string of modules incapacitated as long as some of the modules in the string are shaded. This loss of a full series string can of course be minimized or avoided through the use of bypass diodes, and some shadow-related output losses may have to be tolerated if space is at an absolute premium. However, this should be avoided if at all possible, since only so much of a trade-off may be allowed if the installation is to prove worthwhile.

The sun's seasonal paths for your location can be established by your land surveyor when he lays out your building site. You can also do it yourself if you have access to a transit or wish to purchase a device called a sun-angle calculator. One of the better-known sun-angle calculators is called the Solar Pathfinder. It features clear acrylic sheets with curves representing the sun's path through the sky during different seasons for different latitudes. To find the sun's path at your building site, you select the proper sun path sheet, set the unit level and face it toward true south. You can then view the sun path superimposed on the horizon of your building site and determine if there are any trees or other objects that will cast a shadow on your photovoltaic array. If there is any doubt about shading, sun path data should be obtained for your chosen location before you install your PV system.

The Stand-Off Mount

The stand-off mount might be described as a rack mount without the rack. That is, the modules in the array stand upon a framework separate from and above the finished conventional roof, but that framework is

PHOTO 9-3

This 2.2-peak-kilowatt PV array was installed in an hour and a half by a crew of four using Mobil Solar's innovative stand-off mounting hardware.

only a few inches high and lies parallel to the roof's own pitch. This is the preferred method for retrofit installations on houses having the proper roof angle and orientation where the photovoltaic modules are not expected to take the place of conventional roofing materials.

For stand-off mounting, support rails (usually aluminum or steel; sometimes wood) are fastened to stand-off supports or lag-bolted directly to the roof. The modules are in turn fastened to the rails. As with the rack mount, weather-sealing of the roof need take place only at the points where the bolts penetrate.

Recent innovative work done at the Massachusetts Institute of Technology's Energy Laboratory has resulted in the design of a stand-off mounting system that uses the photovoltaic module's own frame as the support structure and relies on prefabricated roof jacks to support the modules and provide the desired stand-off height above the roof.

The stand-off mount retains most of the rack mount's ventilation advantages if it is designed to allow adequate free air flow behind the photovoltaic modules (between the roof and the array). Since the hardware requirements are less than they are for the rack mount, costs are somewhat lower. But none of the conventional roofing materials is displaced, and consequently no dollar savings in that area may be balanced against the PV installation costs. In addition, when the conventional roofing requires maintenance, the array may have to be dismantled to provide access for repair or replacement.

The Direct Mount

It is the direct mount concept that begins to present savings in the cost of conventional roofing materials. Economy and simplicity are also possible with the direct mount through the elimination of support framework and mounting rails. The installation is exactly as the term implies—*direct.* The photovoltaic modules are not mounted atop a superfluous layer of shingling but are secured directly to the roof sheathing. In some early prototype designs, the modules themselves—sometimes hexagonal in shape—were overlapped like traditional shingles. But regardless of whether they overlap, PV modules in a direct mount must be installed to serve as the weathering skin of the roof. Adequate caulking or sealant that can withstand wind forces, repeated thermal cycling, and prolonged direct solar exposure is essential to this task.

The concept of direct mounting of residential photovoltaic arrays had the attention of PV researchers in the early 1980s. However, there are some notable disadvantages to this method that have been responsible for its general lack of acceptance.

Since the modules are applied directly to the structural roof sheathing, there is no open space between the module and the roof in which cooling air can circulate. This absence of ventilation can result in direct-sunlight operating temperatures as much as 20 degrees Celsius hotter than those common to other mounting methods, causing a significant decrease in electrical conversion efficiency. In addition, module-to-module interconnections require special flat cables and connectors.

Even more important, access to the electrical connections of the individual modules after installation in a direct-mount array is very difficult, making routine diagnostic and maintenance operations a major task. General Electric and ARCO Solar both produced prototype photovoltaic modules designed especially for direct mounting. These designs proved unworkable for the reasons just stated. Currently, no PV manufacturer produces modules that are designed for direct mounting.

The Integral Mount

The integral mount is favored to emerge as the preferred installation method for new residential and light commercial construction where the roof is designed at the proper orientation. It represents a logical progression from the direct mount, eliminating all of its bad features while improving on the module-cooling characteristics of the rack mount. This approach not only eliminates the exterior roofing material, it also gets rid of the wooden sheathing and roofing felt by attaching the photovoltaic modules directly to the building's rafters, making them both the visual and structural surface of the roof as well as its barrier against the weather.

The tempered glass, generally in a 3/16-inch thickness, that is used as the structural element and top "cover plate" for nearly all photovoltaic modules is more than strong enough to serve as a building's roofing material if the array mounting is properly designed. If you are inclined to be skeptical, consider how many stores, shopping centers, and commercial office buildings now feature sloped glass entrance roofs or interior glazed atriums. These spaces most often use the same type of glass found in PV modules, and the spans employed are also about the same. In addition, the PV module is actually stronger than the simple glass sheet because its lamination of many layers of plastic film provides reinforcement against excessive deflection.

Mechanical load standards dictated by building codes, of course, apply to all structures, be they glazed sunspaces, photovoltaic roofs, or conventional construction. However, many manufacturers of PV modules have already seen to it that their products meet or exceed these standards. For example, Mobil Solar's large-area, 4-foot by 6-foot Ra-180 PV module has passed the rigorous mechanical load testing routine performed by Cal Tech's Jet Propulsion Laboratory. It withstood the equivalent of 125-mile-per-hour positive and negative wind loading in an integral mounting configuration without damage or even any leakage.

The effectiveness of the weather seal afforded by a roof in which an array of separate glass panels replaces the several layers of wood, tarred felt, and shingles is obviously essential to the success of the concept. In our designs, the waterproofing is accomplished with the mechanical closure and support. The vertical joints between adjacent modules lie on and are supported by the top of a rafter whose spacing coincides with the 4-foot on-center module dimension. A 2-inch strip of butyl rubber glazing tape is used above the modules to provide the weather seal, and this is capped by a batten of aluminum or wood that is screwed into the rafter.

On arrays that employ just one horizontal row of modules, the horizontal waterproofing is accomplished at the top with a Z section of fabricated cap flashing that covers the top edge of the array and in turn either caps the roof or allows the shingles above the array to be laid over it. When the array is designed with two or more modules in vertical rows, an aluminum extrusion provides module support and watertight closure in the horizontal direction. A strip of soft silicone sponge rubber is used as a gasket and the joint is then sealed with a thin bead of clear silicone sealant after the assembly is complete. The base of the array is finished with a fabricated flashing that either interfaces with the conventional shingles below the array or forms the roof's drip edge.

The soffit at the base of the space directly beneath the array is finished with a ventilation screen to allow an easy upflow of fresh cooling air, and the ridge is designed with a louver to provide an escape route for hot air in the attic above the living area of the house. This space is vented by passive cooling (natural convection), conventional attic fans, or in hot, dry climates, the discharge air from evaporative coolers. With an integrally mounted photovoltaic array, insulation for the house is placed in the space between the living area ceiling and the attic floor, not in the rafter space directly beneath the modules.

One immediate benefit of integral mounting is improved ventilation, which moderates temperatures and assures more efficient operation of the modules. An extra added advantage of this approach is the cooling of the attic space, which results in a more comfortable living environment with less burden on climate-conditioning equipment and less load on the photovoltaic system as well.

The integral mount shares with the rack mount the advantages of allowing you to work from a flat surface during array installation and provides easy access to the electrical connections at the underside of the array out of the weather for initial connection and for future testing and maintenance. Replacement of individual modules within an array is extremely uncommon, but should it ever become necessary you will appreciate this "back door."

Integral mounting of residential arrays holds an appeal beyond that represented by the cost savings in roofing materials or the ease of installation and service. This appeal is partly aesthetic and perhaps partly philosophical and relates to the fundamental departure from traditional concepts of energy supply and building design of which the solar electric house is a part. While the other mounting techniques are unquestionably serviceable and even desirable in the case of retrofits to existing buildings, they nevertheless suggest a certain redundancy: if the modules can serve as the finished roof, why shouldn't they? The promise of photovoltaics is in many ways the promise of a new architecture, and the physical lightness, simplicity, and translucence of an integral array can only reinforce this spirit, creating an energy-producing structure where the building's own skin produces all the energy it requires, with perhaps a surplus for export to less energy-conscious neighbors.

Mounting Balance-of-System Components

The same attention to planning details and logical integration required in the installation of the photovoltaic array are necessary in mounting the balance-of-system (BOS) components. The specific considerations are as varied as the functions of these components: an inverter must be accessible for possible servicing yet out of the way of household traffic; an auxiliary generator should be located where the noise of its periodic operation will not present an annoyance and near a safe storage place for fuel; batteries must be well ventilated yet kept warm with access provided for maintenance. In short, the most trouble-free system is the system that has been well planned ahead of time and put together properly from the start.

Diodes, String Combiners, and Lightning Protection

Following the topology of the photovoltaic installation as the electricity flows from array to loads, the first system components we encounter are bypass diodes, lightning protection devices, blocking diodes, and the string combiners that make parallel connections of the array's module series strings.

Lightning protection, as we learned in chapter 4, is most commonly accomplished by the inclusion of surge arrestors between the positive DC output conductors from the array strings and earth ground. Their installation is relatively uncomplicated. The type of arrestor generally recommended for photovoltaic use is the nonlinear resistor, or varistor, a button-shaped, solid-state device made of silicon carbide, selenium, or metal oxides. The proper installation and use of these devices is covered in Article 280 of the National Electrical Code (NEC), which includes the following stipulations:

- Surge arrestors may be located indoors or outdoors, but must be protected from the weather in an enclosure with a secure cover to prevent tampering.
- The system's main earth-ground connections must not be made through a surge arrestor, although the location of the arrestor is permitted between any conductors and earth ground.
- In order to prevent excessive impedance to electrical surges associated with lightning strikes, the conductor leads on both the ground and conductor sides of a surge arrestor should be as short and straight as possible.

If a string combiner is purchased for the system, the required lightning protection devices will very likely be included as an integral part of the component's circuitry. On smaller photovoltaic systems that don't employ a string combiner, the surge arrestors are often furnished as part of the internal workings of the voltage regulator. (Figure 9-1 shows the proper connection of surge arrestors.)

The bypass and blocking diodes described in chapter 4 should at this stage of photovoltaic module development present little difficulty for the installer of a residential PV system. Most module manufacturers now build bypass diode protection into their products, either internally at intervals between series strings of cells or in the module's junction box. If your array is to be wired with strings of more than two modules in series, bypass diodes are recommended. They should be sized to safely handle maximum design module short-circuit current.

If the photovoltaic modules you have selected do not come with bypass diode protection, you should install the diodes across the output terminals in the junction box of each module. Bypass diodes are installed in parallel with each module, with the diode's anode on the module's negative terminal and the cathode on the module's positive terminal. Diodes should be connected to the module's output terminal posts using crimp-on spade lugs.

Blocking diodes are used to prevent a reverse flow of current into the array strings. They are located in series with the positive DC power lead from each array string, with a diode's anode connected to the positive array terminal and its cathode to the load. (The installation of a blocking diode is shown in figure 9-1.) Blocking diodes should be sized to handle maximum design short-circuit module or string current and to tolerate voltages with a safety margin higher than that of the string when it is in the open-circuit mode. On many photovoltaic systems, the blocking diodes will require a heat-sink mounting. Since blocking diodes are now frequently found as integrated components in factory-made string combiners, they will not usually require separate installation.

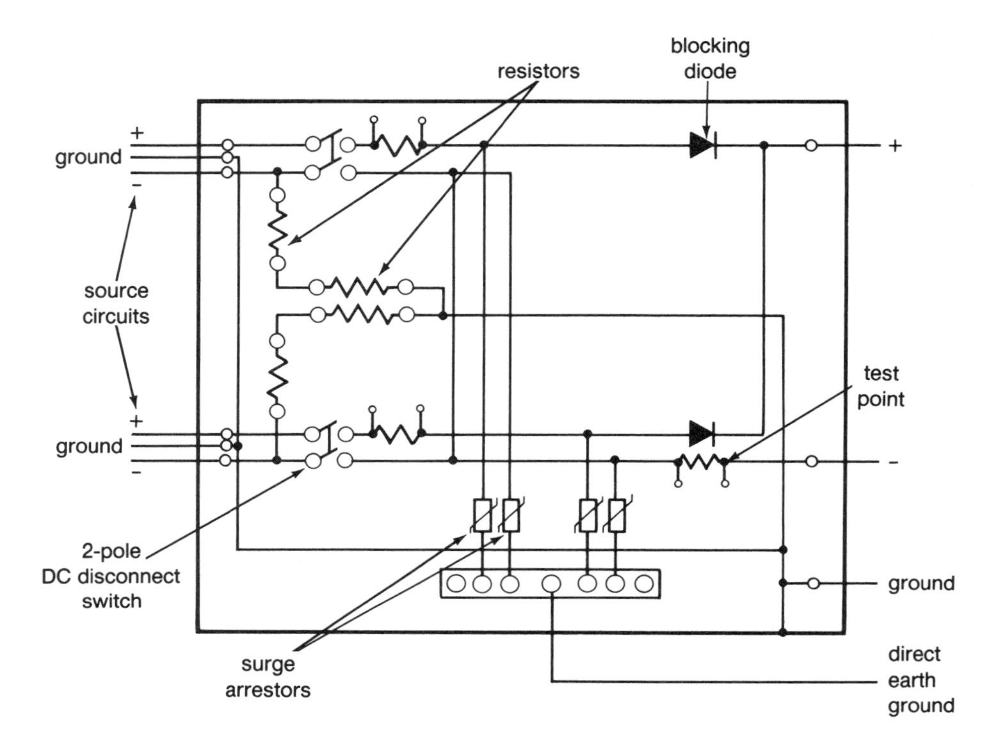

FIGURE 9-1
A string combiner wiring diagram, showing blocking diodes, surge arrestors, DC disconnect switch, and ground wires.

In chapter 4 we saw that small, low-voltage SA photovoltaic systems with all modules wired in parallel will not require any blocking diodes when a series voltage regulator is used. However, if your array is larger and has multiple module strings, blocking diodes should be included in each string even if a series regulator is used, because a fault current, such as a ground fault, which would not be protected by the regulator, can occur between individual strings.

The string combiner is the focal point for input from the series-connected strings of photovoltaic modules. In the combiner, the series strings are combined in parallel to produce the desired array input amperage. A well-designed string combiner will feature heavy-duty terminal blocks for DC array string input and the DC outputs and grounds. It will include DC-rated disconnect switches to isolate each individual string. The current-handling capacity of the DC disconnects should be at least equivalent to the output current of one module, since the module series string current will be the same.

Blocking diodes and lightning surge arrestors of the appropriate rating are often included in the string combiner. However, on large arrays the lightning-protection devices will likely be housed in a separate enclosure located between the string combiner and the array. Capacity for expansion to include additional module strings might also be provided for in the string combiner to make future system upgrading easier.

Once exclusively a custom-built unit, the string combiner is now increasingly available as an off-the-shelf component designed to comply with code requirements, and it is commonly assembled with ease of installation in mind. Photovoltaic component suppliers such as Photron, Inc., and Balance of Systems Specialists can provide string combiners to suit your system requirements.

The string combiner should be installed indoors, near the power conditioner and utility distribution panel in a UI system or near the regulator and battery bank in an SA application. As with all other electrical components, the location chosen should remain dry and be out of reach of children and the casually curious visitor.

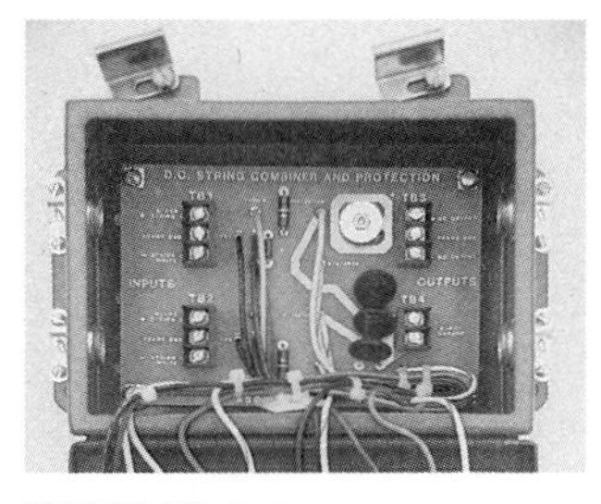

PHOTO 9-4
The interior of a string combiner unit with three surge arrestors (lower center right) and blocking diode (upper center right).

Utility-Interactive BOS Installation

In a UI system, after the photovoltaic array source circuits have been paralleled within the string combiner, the DC output circuit is usually run directly from the array to the power conditioner. Some UI inverters will require DC-input filters that are physically separate from the inverter, and occasionally separate isolation transformers on the output (AC) side as well. However, most units now on the market have been refined to the point where all the necessary circuitry is contained within the power-conditioning unit (PCU) itself.

When mounting the UI inverter, you should follow the same principles that are described for the installation of SA units a little later in this chapter. All the concerns about water, moisture, temperature, and cooling air mentioned there apply here as well. Electromagnetic interference (EMI), radio-frequency interference (RFI), and audible noise often cause problems with UI inverters, although fortunately less so than with SA units. Still, the same preparation and precautions recommended for locating SA inverters should also be followed for UI systems. After the PCU, the PV system output (now AC) is run directly to the standard house distribution panel.

Stand-Alone BOS Installation

After the string combiner, the system installation procedures for SA systems differ from those of UI systems. In an SA system, the photovoltaic array DC output circuit is run to the voltage regulator. The regulator should be installed near the battery bank. This is important for several reasons. First, the regulator should operate at a temperature as close to that of the battery bank as possible. Regulators with remote temperature compensation will need to be mounted within the reach of their remote temperature probes. Second, the DC lines should always be kept as short as is practical to avoid line losses. Third, in small systems where the regulator serves as the system's master controller—managing the battery bank output in addition to the array charging—the distance away from the battery bank is even more important, since the power demand from the load often requires heavier current flow than that from the array.

When mounting the regulator close to the battery bank, be certain that no risk of explosion of hydrogen gas is created. Remember to install your battery bank with adequate provision for venting any gas produced during charging, and never install the regulator with the batteries in a sealed enclosure such as the outdoor battery box sometimes used in small remote photovoltaic systems. The only exception to this would be if the regulator were an all solid-state unit carefully chosen for that application, since in such a unit no sparks from opening relay contacts would occur.

When choosing a mounting location for the regulator, remember that all regulators have an operating temperature range. Be sure that the winter low or the summer high at your site will not exceed this range. Also, remember that the regulator must be easy to wire during the initial installation, easily accessible for checkups and maintenance, and protected from getting wet. Always label your wire runs and the individual conductors, and be sure to maintain your color code to ensure correct polarity throughout the system.

In a small system the regulator will probably just be fastened to the wall in a convenient location, whereas in a larger system with more sophisticated controls, a special panel board will likely be made and (in many cases) the system's charge controller, load-management devices, disconnects, overcurrent protection, and even load distribution will all be prewired in the shop and delivered to the site in a completed assembly. In any case, the principles outlined above apply regardless of how sophisticated the system is.

Battery Installation

The DC array output now goes from the regulator to the battery bank. In a new home built with SA photovoltaics in mind, the location of the battery bank should be determined as part of the initial design process. If the home is to be provided with a basement, one corner can be given over to the batteries, with or without partitioning. In a house with a slab foundation (that is, without a basement) a utility room can be designed to provide unobtrusive storage for the batteries.

For retrofit SA installations in existing homes, the question of battery location may be a little more tricky. We should remember, though, that the vast majority of such systems will be small and will consequently have lighter loads and a small battery storage requirement. Likewise, most such structures are likely to be located in places where there is sufficient room for a small shedlike addition, should one prove necessary.

If the system is a small one, battery location should not prove much of a problem. A few batteries placed on the floor or in a warm, dry crawlspace will probably be all right. However, if your system is larger and there are many batteries, careful preparations must be made to house them. Keep in mind that these are large and heavy components. Large battery banks are very large and very heavy. Therefore, if a large battery bank is to be installed, a concrete floor should be planned, and if the batteries are to be stacked, a heavy-duty structural steel rack should be used.

Regardless of the location chosen for your batteries, it is impossible to overstress the need for preventing them from getting too cold or too hot. When it comes to temperature and performance, batteries are the opposite of photovoltaic cells: the colder a battery gets, the lower its capacity for electrical storage. The figures presented in table 9-1 tell the story.

As temperatures rise above 27°C (80°F), battery capacity actually increases above 100 percent of the manufacturer's rated level. But above 30°C (86°F), at 105 percent of rated capacity, there is a danger of shortened battery life unless a lower concentration of electrolyte (less sulfuric acid) is used. Temperatures in excess of 52°C (125°F) should always be avoided. Conversely, an electrolyte with high sulfuric acid content (high state of charge) can improve efficiency under cold conditions, while also helping to prevent freezing. At 85 percent state of charge, for instance, electrolyte freezes at approximately −50°C (−58°F). At 50 percent state of charge, that figure goes up to −20°C (−4°F). At a 10 percent state of charge, a battery will freeze at only −5°C (23°F). In any event, and at any state of charge, every effort should be made to operate batteries as close to the desired 16° to 27°C (60° to 80°F) optimum range.

It makes sense, then, to keep the batteries in an insulated, draft-free part of the house with an ambient temperature comparable to that maintained in the living areas. Also, try to keep all of the batteries at approximately the same temperature. If it is impossible to keep the temperature of the battery storage area in the optimum range, it may be necessary to compensate by proportionally increasing the size of the battery bank.

Another consideration in choosing a location for batteries is the requirement that they not be situated in an area where there is no circulation of air for ventilation. The phenomenon of "gassing" is common to a greater or lesser extent in all liquid-cell batteries and can be safely handled as long as air circulation is assured and frequent overcharging is avoided.

In the Hudson Valley house described in chapter 8, the need for

Table 9-1. Battery Performance in Relation to Temperature

Temperature (°C/°F)	Battery Operating Capacity (%)
27/80	100
16/60	90
4/40	77
−7/20	63
−18/0	49
−29/−20	35
−40/−40	21

ventilation was taken care of by constructing a manifold system over the assembly of batteries, which vented any gas produced to the outdoors. This manifold system avoided the possibility of cold outside ventilation air threatening the temperature stability of the storage area during the cold upstate New York winter. In the Hudson Valley system, unconstricted lengths of ⅜-inch tubing were connected from the vent port in each battery to a central manifold, which was then vented to the outdoors.

Batteries must be kept dry as well as ventilated. Lead-acid batteries may be located directly on the floor, provided the floor can tolerate the possible acid spills. If the batteries are placed directly on a concrete slab and this slab is exposed to frost conditions, rigid insulaton should be installed under the slab before it is poured. This will prevent the batteries from becoming too cold in winter when you need them the most. As a space-saving alternative, batteries may be placed in racks three or four levels high as long as there is enough room to wire and maintain them. Of course, the rack must be built sturdily enough to support the weight, which can be many thousands of pounds in a large system (the battery bank in the Hudson Valley house weighs over 9,000 pounds).

Air circulation around and through the racks is also important to avoid heat buildup during heavy loading and to keep the batteries at an even temperature. In basement installations where any possibility of flooding exists, batteries should be supported above the floor on racks, even if they are only to be one level high. Moisture lingering around the bases of batteries can accelerate corrosion, and flooding can short the system and create a dangerous set of circumstances. Coming home to find your cellar flooded and your high-voltage DC battery bank underwater can ruin your whole day.

Inverter Installation

All but the smallest SA residential photovoltaic systems will probably include a DC-to-AC inverter. The SA inverter should be located close to the battery bank to minimize wire losses and simplify installation and maintenance. Like voltage regulators and other electronics, inverters are negatively affected by water or moisture and operate best if kept at or near room temperature. However, even at room temperature, most inverters will need good air circulation to cool their power semiconductors and internal transformers. Some inverters even have built-in cooling fans to ensure good internal air flow. Care should be taken with all inverters to ensure that the intake and outlet air passages have good access to cooling air or you'll be replacing expensive power transistors much sooner than you might like.

Inverters are heavy because of the type and quantity of metal found in their power transformers. Large models can weigh as much as a couple of hundred pounds, so plan your mounting accordingly. With that amount of weight, the easiest solution would seem to be to just set the inverter right down on the floor. However, if there is any chance of flooding, floor mounting is not recommended. Ideally, it's best to plan the inverter installation so the unit will be wall mounted about 5 feet above the floor. Holding the unit up off the floor keeps it dry and improves the flow of cooling air. Having it at eye level also simplifies wiring and provides easy access for future maintenance. Fortunately, as inverters are improved, their weight will be reduced through the use of high-frequency switching that requires smaller, lighter transformers.

The amount of noise produced varies widely from unit to unit and should be carefully considered before the inverter location is decided. Remember too that inverters can generate troublesome EMI and RFI. This can be even more of a problem than audible noise, since it is more difficult to quantify and plan for in advance.

The best way to deal with such problems is to seek advice from your photovoltaic system designer or supplier. Also, try to visit with some folks who have been living with the type of equipment you are considering and can give you a firsthand report on its performance. The best defense against both audible noise and RFI is, of course, good inverter design. The second is distance. For this and many other reasons, it makes good sense to locate the inverter and other equipment in a stand-alone PV-powered residence as far away from the living area as possible. And, when installing your inverter, make doubly certain that it's solidly grounded. A solid ground will help to disperse inverter noise. Also, it is required by the electrical code.

Generator Installation

The auxiliary generator is the heaviest of the individual BOS components in an SA photovoltaic installation and the one with the most moving parts. It is also the noisiest and the only one to draw on a conventional supply of fuel. All of these things have to be taken into consideration when locating and installing the machine.

Three factors are paramount in siting the auxiliary generator. These are weight, noise, and the need for air intake and exhaust connections. The weight of a generator (a 4-kilowatt diesel model weighs in the vicinity of 500 pounds and an 8.5-kilowatt set 800 pounds, whereas gasoline- and propane-fired units are somewhat lighter) requires that it be mounted on a solid floor surface, preferably a concrete slab at ground level. In order to best distribute the weight and keep the generator set up off the floor in case of flooding, an additional 4-inch-high concrete equipment-mounting pad is recommended, with anchor bolts set in the concrete to receive the set. Many generator sets come factory-equipped with rubber mounts for damping vibration. If these are not included in your set, they should be added at each point of contact with the concrete base. Several types of spring mounts are available that also provide adequate vibration isolation.

As with inverters, the noise problem can best be addressed by distance. Start by siting the generator as far away as is practical from living areas and by making proper provisions for muffling and exhaust outfall. Since the gen-set in a photovoltaic installation is an auxiliary machine, noise will not be as much of a nuisance as in systems entirely dependent upon gen-set operation. Timers are often included in larger stand-alone systems to automatically start and run the generator and to limit its operation to the part of the day when it is least obtrusive to the home's occupants.

Air intake for both combustion and engine cooling and exhaust outfall are crucial to the operation of any internal-combustion engine. The generator will have to be sited so that intake air ducts and exhaust piping runs are not impractically long or complicated. Running more wiring to a generator site is far cheaper than extending ductwork or piping. Other factors to be considered in gen-set siting include dryness and complete assurance against flooding; ease of access for maintenance; and room, within a reasonable distance, for safe storage of fuel.

In addition to locating the generator as far as possible from the living areas, the main line of defense against gen-set noise is a good muffler. Very small gen-sets come from the manufacturer with the exhaust noise suppression built in. Larger units must be site-equipped with a muffler adequately sized not only to reduce engine operation noise but also to accommodate the exhaust output of the machine. This requirement cannot be overstressed. If the muffler is too small, exhaust pressure will back up and the generator will not burn fuel efficiently or function properly. The location and direction of the exhaust outfall are also very important. The exhaust is a major source of noise and its outfall should point away from the living areas.

As we noted above, the gen-set is an internal-combustion device, similar to the engine in your automobile. Internal combustion, of course, involves not only the burning of hydrocarbon fuels but of air as well—or, more specifically, of the oxygen component in air. In addition to requiring air for combustion, generators need constant circulation of air for cooling. Smaller units—those up to about 5 kilowatts—are generally air cooled, but even the larger, water-cooled models require a flow of fan-forced air to dissipate radiator heat.

Air inlets and outlets must be sized to allow sufficient free-air circulation for combustion and cooling. If too much heated air is recirculated without fresh outside input, its value as a coolant is quickly reduced. However, unrestricted air flow may be undesirable because of cold outdoor conditions. In such cases, openings can be equipped with louvers and/or exhaust fans operated either manually or thermostatically. Installers should consult manufacturers' guidelines for the air requirements of specific machines. As an example, the Kohler Company recommends air inlets and outlets of 1.25 square feet each for a 5-kilowatt gen-set. Remember that louvers, screens, and/or filters reduce air flow and require a proportional increase in the area of the opening.

Exhaust gases from any internal-combustion device are highly dangerous regardless of the type of fuel used. The exhaust system is the means of safely venting these gases, particularly lethal carbon monoxide, to the outdoors. Exhaust provisions for a gen-set located indoors should be as simple and straightforward as possible. We noted earlier that an insufficiently sized muffler will constrict engine performance; so will exhaust piping that is too long or too narrow or has too many bends and turns.

Two components of a gen-set exhaust line are especially important. One is the inclusion of a flexible section of pipe, generally between the exhaust manifold and the muffler. This will reduce noise transmission and prevent engine vibration from straining and possibly shaking exhaust fittings loose, allowing indoor seepage of dangerous gases. The other is a trap equipped with a drain cock located in a section of pipe placed between the manifold and the muffler and sloped away from the generator. The trap keeps water, a combustion by-product, from trickling back into the engine from the exhaust lines after condensing in the exhaust piping.

Gen-sets selected for use with residential photovoltaic systems will be fueled by diesel, propane (or natural gas), or gasoline, with the first two choices by far the most likely. Principal considerations here include the safe and practical installation of storage tanks and fuel distribution lines.

Diesel fuel may usually be stored indoors in large quantities—it is, after all, similar to the fuel used in home heating systems. Storage may be at the level of the generator or overhead, but when the main tank is 5 feet or more above the generator's fuel intake, and/or the horizontal line from the tank to the generator is longer than 6 feet, an auxiliary fuel pump and transfer tank may be necessary for proper fuel flow. Regardless of the location of the tank, a fuel oil filter is required, and its element should be changed regularly. As a precaution against clogs in fuel lines or injectors, use black iron, steel, or copper pipe for the fuel lines; galvanized surfaces react with diesel fuel and deteriorate. Here, too, a flexible section of pipe will prevent loosening and leakage due to vibration.

As for the fuel itself, by far the most common is No. 2-D diesel fuel, as distinguished from ordinary No. 2 furnace diesel (fuel oil), which should not be used in engines. Above 5,000 feet or below 4°C (40° F), No. 1-D, a class that includes kerosene, may give better performance, provided it is blended with lubricating oil at a ratio of 1 quart of oil to every 100 gallons of diesel fuel. The manufacturer's recommendations for your particular generator should always be closely followed.

The storage of liquid petroleum gas (propane, butane, or a mixture of these gases in proportions necessary to maintain pressure at a given temperature) is already familiar to many rural and suburban homeowners who rely on it for cooking or heating. It is stored in a heavy tank either above or below ground. High pressure maintains the contents in a liquefied form except for the vapors that collect at the top of the tank and flow into the fuel lines by way of two regulators. One regulator controls pressure going into the line at the head of the tank; the other releases the gas needed into the engine's carburetor.

There must also be an automatic fuel shutoff so that the regulated flow out of the tank stops when the generator automatically shuts off. The fuel line between the tank and generator also helps to regulate pressure and consequently must be the right diameter. Pipes should be black iron or copper, and once again, a flexible section near the generator is a must. If engine vibration loosens a fitting in a diesel line, leaking fuel will tell the story. Gasified fuels are less forgiving. Vibration-dampening mounts and adequate ventilation are thus even more important.

As we learned in chapter 5, all but the smallest generators are equipped with internal safety controls that will protect the unit from damage during operation from things such as low oil pressure, loss of cooling, ignition failure on start-up, and so on. However, there are two major safety considerations in generator installation and operation that are external to the unit itself. These involve the exhaust lines and fuel supply. Of principal concern is the proximity of exhaust and fuel components. The exhaust apparatus of a generator runs hot and should *never* be close to fuel-bearing lines, *especially* lines carrying gasoline. Gasoline is the most volatile of all engine fuels. For this reason it is the least commonly used of the major fuels for domestic power backup applications.

The heat of exhaust pipes also prompts another concern: if exhaust lines must pass through combustible walls or roofs, the same precaution must be taken as with hot stovepipes. This involves the use of insulated "thimble" fittings at points of penetration. The local building and fire codes should always be consulted first for specific regulations with regard to construction of a generator enclosure and the storage and piping of fuel.

Battery Chargers

The location and installation of an AC-to-DC battery charger follows the same general principles as the location and installation of an inverter. Battery chargers have the same requirements with regard to temperature, water, and cooling air. Like some inverters, most battery chargers feature an internal cooling fan that must be free to provide unimpeded flow of cooling air through the unit. Like inverters, battery chargers are heavy and floor mounting would seem an attractive option. However, floor mounting of battery chargers should be avoided unless you are certain that there is no chance of the unit getting wet. Mount the battery charger as close as you can to the battery bank to minimize the length of the wire runs and be sure to leave room for disconnect switches on both the AC and DC sides of the unit.

A guiding theme of this chapter has been that a photovoltaic installation should be viewed not as a collection of discrete units but as an integrated system. As such, each stage of the installation—from selection and mounting of modules to selection and positioning of BOS components—should be carefully considered in relation to the entire system. Careful advance planning is necessary to ensure that every decision made regarding equipment and its interconnection will minimize difficulties that can arise in the installation and operation of a PV system.

Safety and convenience should both be kept constantly in mind in the decision-making process. For example, the inverter should be located close to the batteries in a place easily accessible to a technician but should be kept well away from water and from children. Likewise, the inverter and the generator should be located where their noise and electronic interference cause minimal disturbance to the residents of the home and neighboring homes. Practical considerations such as these are an important part of living in the solar electric house.

This chapter began the discussion of the installation process. The chapter that follows completes the discussion by examining the vital topic of the proper methods and materials to use when wiring a photovoltaic system and concludes with instructions on how to properly operate and maintain the system once it has been installed.

CHAPTER TEN

PHOTOVOLTAIC SYSTEM WIRING AND MAINTENANCE

At the beginning of chapter 3 we saw how cells are joined together to form modules and modules are linked together to make an array. There we noted that the way to increase the voltage (that is, the electrical force or pressure) of solar cells is to link them together in series. This is done by running a wire between the back (positive) contact of one cell and the front (negative) contact of another. By contrast, amperage, or current flow, is increased by linking cells in parallel—that is, running wires from front contact to front contact and back contact to back contact. An individual module may have cells joined in either or both ways.

As we proceeded through the steps involved in moving from cell to module and module to array, we learned that the basic topology of a residential photovoltaic array calls for the series connection of individual modules into series strings that deliver the required voltage, and the subsequent connection of these strings in parallel in order to arrive at the necessary output amperage.

The advantages of this approach to intermodule connection over other methods—such as parallel-first or a combination of series and parallel—include lower costs (less wiring material and labor are involved), easier isolation of an individual string for the purpose of module replacement, easier troubleshooting, and fewer complications in the event of future array expansion. The parallel-first method is also less desirable because it requires the use of much larger bypass diodes, which have to be sized to handle the output current of the full subarray.

Keep in mind, though, that series-string array assembly calls for the connection of a bypass diode in parallel with each module. As noted previously, these diodes are often provided with the photovoltaic modules from the factory. It is important to note that these factory-supplied diodes are not able to handle the current levels of a parallel-first wiring arrangement.

The first step in array wiring, then, is module-to-module series connection. The actual points of contact will be the junction boxes on the back of each module (we are assuming an installation incorporating commercially available modules). According to the National Electrical Code (NEC), these connections must remain accessible after final installation of the modules and series string. This will be no problem as long as the modules are held in place with removable hardware or, as is the case with rack-mounted and integrally mounted roof installations, are accessible from behind. With this requirement accounted for, the next choices involve the type of wire used and the methods used for securing wire to modules' output terminals.

Code-Approved Wiring

The NEC has long addressed questions regarding the proper wiring for specific circumstances and has included provisions for photovoltaic (PV) systems for the first time as Article 690 in its 1984 edition. (This edition is listed in the bibliography.) The NEC stipulates that wiring be capable of safe conductivity for the service it will see in either wet (outdoor) or dry (indoor) circumstances and specifies the types of wire that are acceptable in each situation. The PV array wiring that is exposed to the outdoors has the most demanding requirements and warrants special attention.

At present there is only one method for exposed outdoor array wiring without a protective conduit that is code-approved specifically for photovoltaic applications. This is Type UF single conductor cable. Type UF is designed for direct burial and is available with either single or multiple conductors sized from #14 (copper) to #4/0. It is currently the only wire type specifically cited in NEC Article 690 as acceptable for use in module string interconnections. It also must meet the requirement set forth in NEC Article 339 that all wiring to be directly exposed to the rays of the sun be the type designated "sunlight resistant" by its manufacturer.

My firm has had extensive experience using single-conductor Type UF cable for photovoltaic array string wiring and the results have been excellent. Unfortunately, this product may be somewhat difficult for PV installation tradesmen to obtain in certain areas, although some PV suppliers are now beginning to stock it.

Recognizing the potential problem for the industry of there being only one code-approved method for photovoltaic array wiring with nonshielded conductors, the U.S. Department of Energy commissioned Underwriters Laboratories (UL) to review all of the presently available wiring methods, both shielded and nonshielded, and make recommendations as to which might be suitable candidates for code approval in PV array installations. The results of their investigation were published as *Development of Photovoltaic Array and Module Safety Requirements* (Underwriters Laboratories, 1982). Some of the most promising candidates are discussed here.

Type SE is a single-conductor or multiconductor, double-insulated cable that is also sunlight resistant and suitable for direct burial (Type USE). Type SE is available in conductor sizes from #12 (copper) through #4/0 and larger and is suitable for service entrance and branch circuit wiring as set forth in NEC Article 338. Underwriters Laboratory concluded that Type SE and USE were "attractive for photovoltaic installations" and noted that "there were no use-restrictions [in the NEC] which would preclude its acceptance in PV applications by the inspection authorities." Single-conductor Type SE is reported to be somewhat easier to obtain than single-conductor Type UF; however, it is not specifically code-approved for PV applications as of this writing.

Type TC is a multiconductor, double-insulated cable that is available in sunlight- and corrosion-resistant forms and is suitable for direct burial. Type TC cable contains either Type THHN or THWN conductors in sizes starting from #18 (copper). It is suitable for power or control wiring, is readily available, and is reasonable in cost. Underwriters Laboratories felt that Type TC would appear to be a good choice for photovoltaic module interconnections. The UL report went on to cite present code restrictions regarding support of Type TC cables that would need to be revised before Type TC could be code-approved in PV applications. There are many cases where Type TC has already been used to wire PV arrays. However, it is not yet approved by the NEC for this purpose.

Types SJ, SJO, SO, and SEO are double-insulated, multiconductor flexible cords that have been widely used for module-to-module interconnections, although their outer jacket is not specifically rated for outdoor use unless designated "W-A." Underwriters Laboratory found that cables of these types have many attributes favorable to their use in photovoltaic applications. However, they went on to say that further development and qualification testing of these cables would be required—especially in regard to wet service and ability to withstand physical abuse—before they could be considered for code approval in exposed PV array wiring applications.

PHOTO 10-1

A rear view of a rack-mounted PV array, showing module-to-module connections made between individual junction boxes with Type UF single-conductor cable and Heyco weathertight connectors.

Photovoltaic system wiring that is enclosed in a conduit or raceway can employ "conventional" conductors such as THWN, THHN, or TW. And indoor residential PV system wiring can be run with the common Type NM nonmetallic sheathed cable (Romex) that is used for virtually all household branch circuit wiring. When specifying conductors for DC circuits, the color code to observe is positive is black, negative is white, and ground is green. If a PV system is to be designed with three DC power conductors—as with a center-tapped array—red and black should be used as the positive power conductors and white as the negative or neutral, with a green fourth conductor functioning as frame ground.

This color code is also the same in AC circuits, where black is always the "hot," or active, conductor and white is always the "neutral," or ground potential, conductor. In three-wire AC circuits—such as a 240-VAC single-phase residential service—the red, black, and white color code applies, with red and black serving as active power conductors and white

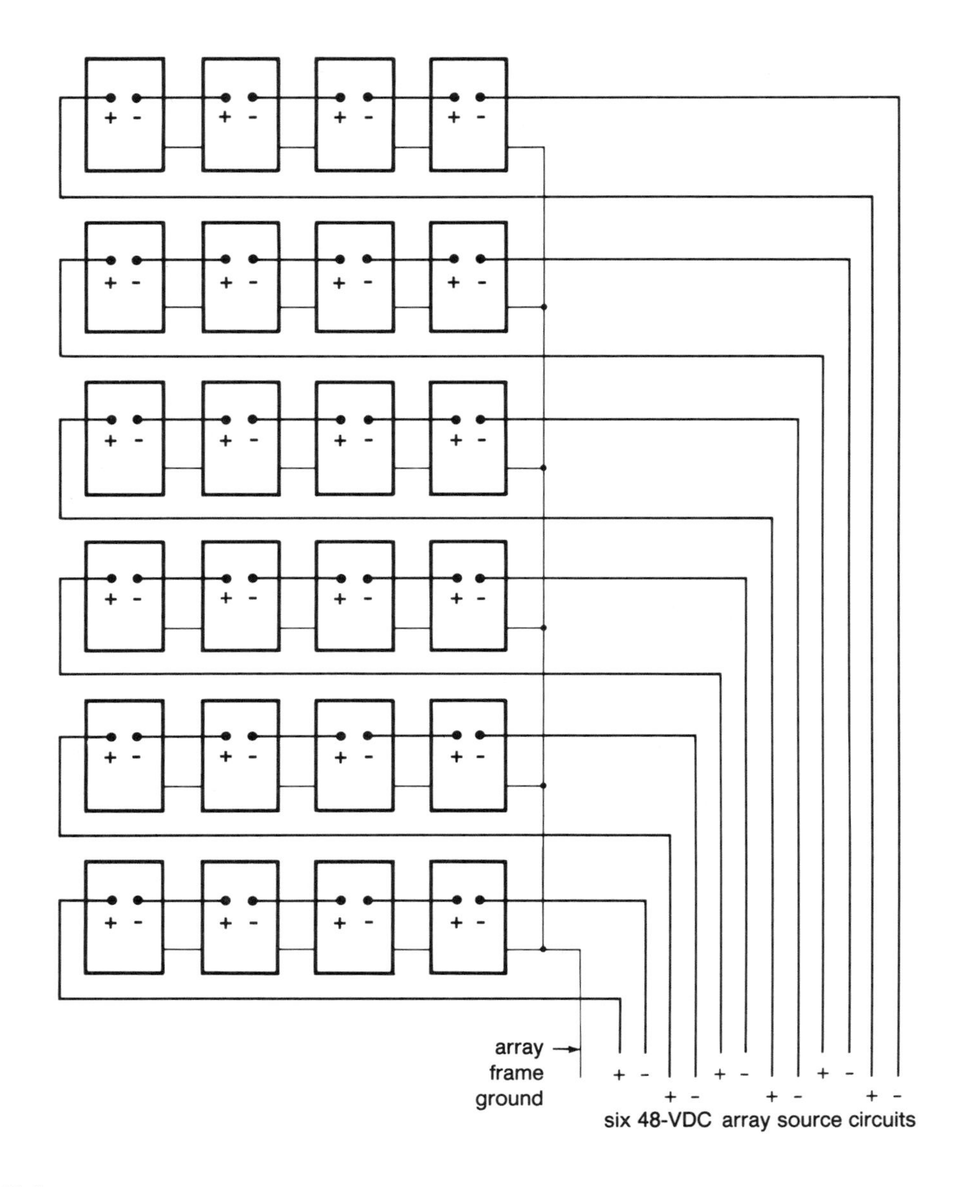

FIGURE 10-1

An array wiring schematic of the stand-alone system described in the second sizing example in chapter 3. Twenty-four 12-volt, 30-watt PV modules are wired in six parallel strings of four series modules each to deliver 720 peak watts at 48 VDC. The six series strings are run from the array to a through-the-roof junction box and then down to the lightning protection and string combiner. Note the array-frame ground required by the NEC. Utility-interactive arrays are wired in the same manner; however, they often have many more modules in series to deliver the higher DC voltages required by line-tied inverters.

as the neutral. Even though AC polarity changes 120 times per second, it is every bit as important to keep the color coding of conductors consistent in AC circuits as it is in DC circuits.

It is important to note that the NEC requires solid grounding for all exposed non-current-conducting metal parts of a photovoltaic system, such as module frames, array support structures, and equipment enclosures. It is recommended that a separate grounding conductor of the same wire type and size as the DC power conductors be run from the array down to the string combiner and from there to earth ground. All individual module frames should be solidly grounded together by means of ground jumpers or a common metal support structure. A wiring schematic for a typical roof-mounted PV array is shown in figure 10-1.

Sizing of the wire used in module-to-module series connections must be based upon the wire's rated ampacity as it relates to the normal operating current levels in the array. Stated simply, the wire has to be capable of handling the maximum amperage produced by a single module if it is to be used only for series connection, since this method of combining module output does not result in an amperage boost. Ideally, some margin of safety in rated ampacity is desirable. (The current-carrying capacities of different gauges of wire are listed in table 10-1.)

Table 10-1. Resistance and Ampacity Characteristics of Type THHN Insulated Wire

Wire Gauge	Resistance (Ohms/100 Ft @ 68°F)	Maximum Recommended Current (A)
14	0.253	15
12	0.159	20
10	0.100	30
8	0.063	55
6	0.040	75
4	0.025	95
2	0.016	130
0	0.010	170
00	0.008	195
000	0.006	225
0000	0.005	260

SOURCE: Article 310 of the 1984 NEC, Table 310-16.

Connectors

Module-to-module interconnections should be made with enough slack to permit the removal of any single module from the array without difficulty and should be secured to the module frames or support structure with wire ties or clips. Connection to the module junction box should be made with either a code-approved plug/receptacle connector combination or a weathertight, through-the-wall junction box connector. AMP, Inc.'s Solar-Lok is one high-quality, code-approved plug/receptacle system. Heyco's pressure-type, liquid-tight gasket connectors are a good through-the-wall alternative.

The AMP connectors are very convenient and facilitate easy array assembly and module replacement. However, they are rather expensive and require a special connector installation tool that would not usually be available to electricians in the field. The AMP system's easy assembly can be taken advantage of if the array wiring can be carefully laid out in advance and the module interconnect wiring can be prefabricated in a shop that has the special connector tool. The Heyco connectors require a bit more field labor but are probably less costly overall, depending on labor rates and the size of the project.

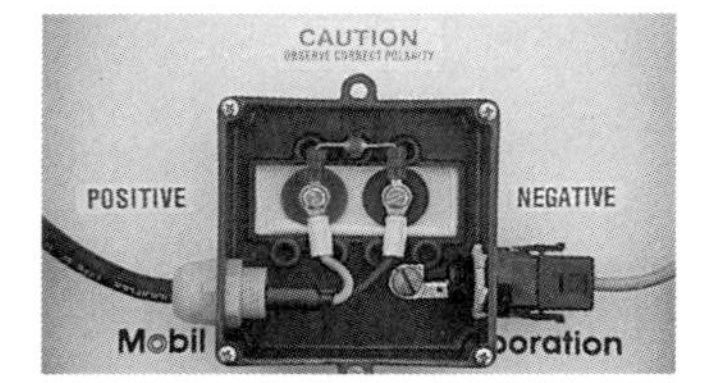

PHOTO 10-2

The interior of a typical module junction box, showing electrical output terminals and bypass diode.

Now comes the task of connecting the string conductors to the module's output terminals. Possible connection options include pressure cable connectors, wire binding screws, and stud-and-nut terminals. Which option is selected will be largely determined by the hardware supplied by the module manufacturer. Mobil Solar provides AMP Solar-Lok quick-connect interconnects as standard items with their Ra-180 modules and as options on their smaller units. ARCO Solar has recently introduced a new quick-connect junction box that requires no additional hardware for wiring.

Crimp-on spade lugs are recommended for use with the terminal-post-type connections that are still furnished on the majority of modules. The junction box should be provided with a weathertight gasket, and all junction box connections, including the through-the-roof penetration, should be accessible after array installation by a simple removal of fasteners. Array wiring connections that are not readily accessible could be considered concealed. For this reason, they may not be allowed by a wiring inspector.

Through-the-Roof Wiring

As was mentioned before, there is an important difference between the photovoltaic system wiring that is above the roof and exposed to outdoor conditions and that which is inside the building and protected from the elements. In designing PV system array wiring, we have found it desirable to make the transition between outdoor and indoor wiring in as short a distance as possible. In commercial arrays, we put each source circuit immediately into a string junction box and then into conduit. In residential systems, we lay out the array to include a through-the-roof junction box for every two strings, which includes four power conductors plus an array frame ground.

This junction box serves both as the point of transition between outdoor and indoor wiring methods and as the entrance for the photovoltaic array wiring into the house. The assembly consists of a weatherproof exterior junction box placed upon the roof surface and sealed, a short length of threaded electrical conduit that passes through the roof, and an interior junction box attached to the other end in the attic space. Array string connections are made inside the weatherproof box with AMP Solar-Lok or Heyco connectors. The remaining system wiring can be done with indoor-rated conductors. This assembly is inexpensive, easily installed, and provides a code-approved transition from the special wiring rated for exterior use to the less costly Type NM Romex.

PHOTO 10-3

Silicone sealant is applied to a through-the-roof junction box assembly for the stand-alone PV system at the Adirondack Mountain Club's John's Brook Lodge.

A word of caution for those installing and wiring photovoltaic arrays: It is very important to remember that PV modules generate electricity directly from the sun *even when the system is only partially installed.* Unless you work nights or install a protective covering over the array, all of the electrical connections you make between modules and between array strings are live. When you are dealing with the high voltages present in utility-interactive (UI) and large stand-alone (SA) systems, a mistake could easily be lethal. It's very easy to forget this, and even if the shock that results doesn't seriously injure you, the fall from the roof probably will. Therefore, it is recommended that PV array and string wiring be done only by electricians or other qualified mechanics trained in working with live conductors. This is especially important when working with larger arrays and higher voltages.

One way of reducing the risk of working with live conductors is to wire the array strings with connectors that plug into the rooftop string junction box(es) and then leave them unconnected while the remaining wiring is being installed and connected from the roof to the string combiner. After the string combiner has been installed and connected, it can serve as the means to disconnect the live conductors coming in from the array.

Wiring the String Combiner

When it comes to wiring the array up to the string combiner and connecting with it, we are still operating within the framework of the NEC rules on intermodule DC connections. We must keep in mind that rated ampacity of the chosen conductors must at least equal (and, to avoid resistive losses, preferably exceed) the amperage output of each series string (called a *source circuit* in the NEC). When the strings are combined in parallel, amperage is increased, and conductors within the string combiner carrying more than the output of one string will have to be sized to accommodate this larger current.

The positive and negative DC-power conductors from each array string are brought in to the string combiner box along with the array frame ground conductor using code-approved wire connectors and are fastened to the input terminal blocks using crimp-on spade lug connectors. The NEC requires that one side of the DC array source circuit be solidly grounded. (The code allows for possible exceptions to this requirement if the alternate means affords the same level of safety.) It is recommended that the negative side of the PV array source circuits be connected to the system's ground bus as shown in figure 9-1. It is important to note that this provision requires that the power conditioner in a UI photovoltaic system be designed with an isolation transformer. The vast majority of UI power conditioners are now designed with isolation transformers, so this should not present a problem.

A heavy-gauge insulated wire, usually #10 or greater but always at least as large as the string power conductors, is directly connected to an earth ground (driven ground rod or copper water pipe) from the appropriate terminal. This direct earth ground is required to provide a direct path to ground for any high-voltage transients diverted by the lightning protection devices that might otherwise damage sensitive electronics downstream in the system. Article 250 of the NEC has specific requirements for equipment and circuit grounding and should be consulted prior to installation. The path of the connection to earth ground should be as short and straight as possible. (See the wiring schematic in figure 9-1.)

Blocking Diodes and Lightning Protection

Earlier we noted that blocking diodes, which prevent a reverse flow of current into the photovoltaic array, are often built into commercially available string combiner units. If they are not, they should be installed in the string combiner box and wired in series with the positive conductor of each module string so that the anode of each diode is on the array side and the cathode is on the load side, as shown in figure 9-1. Be certain that the blocking diodes selected have an appropriate current and voltage rating and are heat-sink mounted if necessary.

If you are using a commercially available string combiner, the nonlinear varistor (resistor) described previously will likely already be installed for lightning protection. However, if the array is large and many varistors are required, a separate enclosure may be necessary. Lightning protection devices should be installed between all nongrounded conductors and a direct earth ground. Crimp-on spade-lug connectors and terminal blocks are convenient to use and make the replacement of varistors, which is occasionally required, much easier than if they were hard soldered in place. For specific installation guidelines on lightning protection devices, refer to Article 280 of the NEC.

Utility-Interactive Balance-of-System Wiring

The string combiner is the point at which wiring specifications for utility-interactive and stand-alone systems diverge. In a UI system, the next step is to run wires from the string combiner to the power conditioner. At this point, of course, the wiring will be carrying the full array output amperage and must be sized accordingly to avoid resistive losses. Again, the ideal approach is to keep the wire runs short and use a heavier cable (one with a lower gauge number) than is strictly dictated by output amperage. This provides an extra margin of safety and facilitates possible expansion of the system later on. Properly sized conductors are connected to the positive and negative DC-output terminals in the string combiner, along with a chassis ground wire. These three wires are then run to the appropriate terminals in the power conditioner.

The NEC specifies acceptable wire composition and sizes for combiner-to-inverter wiring. Type NM Romex two-conductor with ground of a gauge appropriate to carry the system's maximum output current is often used in residential applications. In a commercial application where conduit is required, Type THHN conductors are often used with it. Indoors it may be enclosed in flexible conduit or EMT tubing, whereas outdoors and in wet areas it should be run through RMC conduit with threaded fittings.

From the inverter, the system's output power must be carried to the home's main distribution panel. The two AC power output leads from the inverter are wired to a two-pole, 240-volt circuit breaker. The circuit breaker is sized to exceed the photovoltaic system's nominal rated current output by 30 percent and is set inside the house service panel. The ground from the inverter is connected to the neutral/ground bus in the distribution panel. Generally speaking, the wire used to connect the inverter with the service entrance circuit breaker in the main distribution panel will be of the same NEC-approved gauge and construction as that run between the string combiner and inverter.

The code requires that there be a means of disconnecting all ungrounded power conductors from both sides of the inverter and all other major discrete components in a photovoltaic system. The DC-rated disconnect switches for each array string, which are located in the string combiner, will satisfy this requirement. However, NEC Article 230, Part H, states that there must not be more than six separate disconnect switches. If the system has more than six strings, a separate master disconnect switch must be installed on the DC-input side of the inverter after the string combiner. Whatever disconnect means are used, it is recommended that they be located within sight and in close proximity to the inverter.

The 220-volt circuit breaker that serves as the entry point of PV-generated power into the house distribution panel can also satisfy the code requirement for a means of disconnection on the AC side. It should also be located within sight of and in close proximity to the inverter. These requirements are intended to ensure that the disconnects are always visible and easily accessible to someone working on the unit. They also make it very unlikely that the disconnects could mistakenly be turned on by someone else during the servicing operation. If the distribution panel is not within sight and close to the inverter, a separate master AC disconnect switch should be installed.

In addition to these DC and AC disconnects, certain utility companies may require an externally mounted disconnect switch that they may use to ensure that the photovoltaic system is disconnected from utility lines while line service is under way in the area. This requirement was imposed on the first couple of utility-interactive PV-powered houses designed by my firm. That was in the early days, when utility engineers were just getting acquainted with photovoltaics and so took a conservative approach. Since then, the reliability of the automatic disconnect circuitry internal to UI power conditioners has been clearly demonstrated, and in some cases, it has even been given UL ratings. Therefore, the utilities don't often ask for separate disconnects anymore.

Metering

At this point the installer of a UI photovoltaic system will have to consider the options available for metering the flow of electricity to house loads from the outside (that is, from the grid), as well as the flow of surplus PV power into the grid. As we learned in chapter 2, one of the advantages of line-tied PV systems is the opportunity that they offer for sale of this site-generated surplus to the local utility.

Depending on the preference of your local utility, there are two arrangements that can be used to keep track of this credit-and-debit situation. One involves the use of a two-way meter that will run in whichever direction power is flowing at a given time and yield a figure that represents the periodic net. The other calls for the installation of two single-directional ratcheted meters, each of which will record power flow in a different direction. In this case, the net figure is derived from a reading of both meters.

A decision on the metering arrangement to be used must be made in conjunction with the local utility, which will also indicate its preferences as to meter location and connection to the service entrance. In most UI photovoltaic installations, wire runs between the inverter and service entrance/meter panel are brief, with all equipment mounted in proximity for easy access, service, and reading of meters. In conventional situations utility meters are most often mounted outdoors. However, the recommended proximity to power-conditioning equipment and service entrance, as well as the likelihood of more frequent consultation of the meter(s), will generally argue in favor of indoor installation in a utility-interactive PV home—assuming the utility company approves.

The remaining consideration in wiring the indoor balance-of-systems (BOS) segment of a line-tied photovoltaic installation is the service of loads from the main distribution point, or service entrance. In general, the task from this point on becomes the same as that of distributing power to loads in a conventional house. Wiring must be sized with sufficient carrying capacity to prevent voltage drops that could damage loads. For example, a voltage drop of only 5 percent can damage some electric motors and result in premature failure.

Household loads having the heaviest amperage requirements, such as ranges and clothes dryers, will obviously require the lowest-gauge (heaviest) wire and should therefore be located closest to the service entrance. Remember, though, that careful consideration should be given to the decision to include such heavy-draw appliances in a photovoltaic-powered home.

Stand-Alone System Wiring

Wiring the stand-alone photovoltaic system presents a somewhat different set of circumstances, particularly when DC distribution to the house is involved, but the essential requirements of sizing wire according to the amperage to be carried and otherwise meeting NEC stipulations remain the same.

The wiring of modules into series strings and series strings into the string combiner is accomplished in the manner described for a UI installation. The next step, in an SA system, is to run the DC-output circuit from the main array to the voltage regulator, which controls input (charging current) to the storage batteries. Wire of a composition acceptable in similar applications—for instance, the connection of string combiners to inverters in UI systems—is suitable here, provided particular attention is paid to wire size in order to prevent an appreciable voltage drop between the array and regulator.

The same is true for the next step in the installation, making the main power connection from the regulator to the batteries, with the special requirement that all wiring and wiring insulation materials used in proximity to lead-acid batteries be resistant to the corrosive fumes that they can give off.

Battery Bank Wiring

There are many different types of storage batteries that can be used for photovoltaic systems, and their physical and electrical characteristics vary widely. Principal considerations with regard to battery bank wiring are the amperage rating of the individual batteries, as well as the storage bank as a whole, and the type of terminal connections. The installation and wiring of storage batteries is covered under Article 480 of the NEC. The single most important thing to remember when wiring a storage battery bank is that you must use heavy-gauge wire.

A single 12-volt automotive-size storage battery is capable of delivering several hundred amps of current. Most often the battery bank is configured at a relatively low DC voltage when compared to AC service voltages of 115 or 230. Remember that if voltage is low, then amperage must be proportionally higher to deliver a given amount of power (wattage). A 2.5-kilowatt inverter, for example, working on a 24-VDC input, will require a constant supply current of over 100 amps when operating at full load. When load surges are taken into account, the supply current can reach much higher than this for short periods. If there are other DC loads placed on the system that can draw current during inverter operation, this current must also be added to the inverter supply requirements when sizing the battery bank wiring.

The peak battery bank current for this modest 24-volt residential photovoltaic system will probably be in the range of 130 amps. If there are any future plans to increase inverter capacity to (perhaps) 5 kilowatts, another 130 amps should be added to this figure when selecting the conductors for the battery bank wiring. Referring to table 10-1, which lists the ampacity of conductors, we find that an AWG #2 conductor is the smallest legal size that can be used for a 130-amp load and that an AWG #4/0 is the minimum size that can be used for a 260-amp load.

In both these situations, good design practice calls for the use of conductors heavier than the smallest legal size for normal peak loads: at least a #0 conductor in the first instance, and something heavier than #4/0 cable in the second. Since cable larger than #4/0 is difficult to work with, the ideal approach in the latter case would be to increase the voltage of the system so that a lighter-gauge conductor can be employed.

Conductors that are commonly used for battery-to-battery interconnections include Type THHN, Type THW, Type THWN, and Type TW. Of these, all are suitable for use in wet locations except some versions of Type THHN. All come stranded, and often there is a choice in the thickness of the individual strands of wire. When a choice is available, the conservative approach is to choose the wire with the heavier strands—even though installation will be slightly more difficult because the heavier wire will be harder to bend in tight radii. The heavier strands are desirable because they will withstand the corrosive environment and carry current better than the lighter ones. Some manufacturers actually use welding cable for their interconnects. When selecting a conductor for battery bank wiring, choosing one rated as "oil resistant" is recommended because it will also improve the performance in the presence of corrosive vapors.

Many battery suppliers will provide precut "jumpers" for battery-to-battery interconnections. These will be custom-made to suit your system requirements in terms of amperage and length and will be supplied with factory-installed cable terminations that are compatible with the batteries you have chosen.

There are nearly as many different types of battery terminals as there are types of batteries. Among the choices are top, side, post, stud, and flag terminals. If you choose to make up your own battery interconnects, make certain that the cable terminations you use are compatible with your batteries. This is rather simple to do if you wait until the batteries have arrived on-site. However, many installers prefer to prefabricate the battery jumpers in the shop prior to going to the job site because of the advantage of controlled working conditions and the resulting reduction in field labor. In any case, avoid the type of cable termination that relies on pressure from a screw to hold the conductor in place. Select instead a heavy, crimp-on type of lug or a soldered or brazed lug.

If you have a medium-size to large system and your battery bank is configured with several strings of batteries connected in series to deliver the desired system voltage and paralleled to deliver the desired amperage, you will want to install "lattice" wiring. Lattice wiring consists of redundant parallel connections within the battery bank, placed between the individual strings at each battery or perhaps every two batteries. Lattice wiring is done with much lighter conductors (usually #12) than the main current-carrying series connections and is desirable because it serves to distribute the charging current more evenly over the entire battery bank. It is especially effective during the required equalization charging cycles.

It is also good wiring practice to configure the battery bank connections symmetrically so that the current flows as close as possible to the same distance when entering or leaving each individual battery, as shown in figure 8-2.

If you build a support rack for your battery bank using electrically conductive materials (such as steel angle iron), the NEC requires that the support structure be solidly grounded. The code also requires that an insulating material (other than paint) be placed between the support frame and the batteries. This latter requirement is probably a carryover from the days when the majority of batteries were built with metal outer housings and seems somewhat less important now with batteries supplied in plastic or rubber-based molded cases.

A main DC-rated disconnect switch will be required for the battery bank right as the main conductors come off the batteries. Make certain that this switch is DC-rated to handle at least the full current normally expected in the system when switched under load. If your system is large, expect a fairly large and pretty expensive piece of switch gear to fill this requirement.

In addition to the main battery disconnect, some means of "catastrophic" overcurrent protection is required for the battery bank and should be installed in the main positive output power lead from the battery bank as close as physically possible to the point where the individual series strings are connected in parallel. Keep in mind that a single automotive-type battery can develop several hundred amps under short-circuit conditions and the battery bank in the Hudson Valley house can deliver over 10,000 amps! (To help you understand the kind of current we are talking about, take note of the fact that it is possible to electrically weld steel with 30 amps or less.) So the catastrophic fuse is absolutely essential. It will not protect individual pieces of equipment from faults (they should be protected by their own fuses or circuit breakers). What it will do is prevent the house from burning down in the event a short or ground fault occurs in an otherwise unprotected portion of the system.

One manufacturer makes a line of fuses for catastrophic protection of battery banks that are called Amp Traps and are designed to screw directly onto a stud battery terminal. These are nice, but they are rather expensive; a standard Type FRN fuse (such as the Buss Fusetron) of the proper rating will do just as well.

When you order your batteries, be certain to have them shipped filled with electrolyte (shipped "wet"), since handling acid isn't much fun. Also be certain that they are fully charged just before they leave the factory. Even though this service may cost a little more, it is more than worth it. Putting the initial charge into a battery bank of any size without large, high-capacity, grid-connected, industrial battery-charging equipment can take forever and may put unreasonable stress on your home power system equipment.

After you've ordered your batteries delivered to the site fully charged, it is very important to remember that they *are* fully charged. Just like PV modules being installed in the sun, you must be constantly aware when you are handling batteries that they are live. This may sound simplistic, but getting tangled up with live DC power leads from a high-current battery bank isn't very pleasant. When you are handling the batteries and working over them with tools—running wires and connecting terminals—it's very easy to forget and lay a wrench across the top of the bank or start to hand-tighten a positive terminal while contacting the negative side of the bank or ground. Battery banks can pack a tremendous amount of power, and higher-voltage battery banks can deliver fatal shocks.

Because batteries are so dangerous and mistakes are so very easy to make, I recommend that the installation and wiring of batteries be done only by electricians or other qualified tradesmen trained in this work. This is especially important in larger systems with large, high-voltage storage banks.

Generator and Battery Charger Wiring

Wiring the auxiliary generator in a stand-alone photovoltaic system can be either pretty simple or rather complicated, depending on the size of the system and its complexity. The basic tasks involved are these: (1) wiring the generator output to the battery charger and other AC loads; (2) establishing some means of control of the equipment and the transfer of the load from inverter operation to generator operation; and (3) providing disconnects and overcurrent protection for all of the equipment involved.

In a very basic SA system, the generator's AC output is wired to the AC loads and to a battery charger. When you notice that the battery voltage has dropped too low, you shut off the inverter and go out and start the generator. When you see that the battery voltage has gotten back up to an acceptable level, you go back out and shut down the gen-set and then turn the inverter on again.

The wiring for this basic system is pretty straightforward and the required control functions and load transfer are all done manually. The generator's output consists of one or two active, or "hot," wires, a neutral, and a ground wire (sized to handle the maximum current the generator can put out). These wires are run through a manual, double-pole, double-throw transfer switch to the AC load distribution panel. The battery charger is connected to this panel as any other load might be. When the generator is in operation, the battery charger is turned on manually. When the batteries reach the desired level of charge, the battery charger will shut itself down. The owner will have to then shut down the generator and transfer the load back to the inverter.

It is very important to note that inverters don't get along well with generators, since there is no phase synchronization between the two pieces of equipment. It is imperative that the system be wired so that the inverter and the generator cannot be connected to the load at the same time. If they *were* to become connected together, one of them (namely the inverter), and possibly both, would become a smoky mess very quickly.

In this most basic system, we rely on the homeowner to manually turn the individual pieces of equipment on and off as required to transfer the load over and then manually transfer it back. For basic power at a rustic hunting camp, for example, this system will satisfy the essential requirements. However, it lacks the convenience of a system that runs itself, since its successful operation requires that someone be close at hand to watch over it a good part of the time.

The more sophisticated stand-alone system will have automatic controls that are interfaced with the system's charge controller and constantly monitor the battery state of charge. When the battery bank drops below a certain voltage level, the control will start the generator, make certain that it is running, and then see that it is up to operating temperature before automatically transferring the load from the inverter to the generator. The battery charger will be brought on-line with the load transfer to the generator and will begin to charge the batteries. The charger has its own internal control circuitry to determine the proper amount and duration of the charge cycle. When the batteries have been brought back up to their desired level of charge and/or when the sun comes back to deliver array charging output, the master control will shut down the generator and transfer the load back to the inverter. (See the schematic wiring diagram of this system in figure 8-3.)

The selection of conductors for the connection of the generator's output to the transfer switch depends on the location. Type NM Romex is code-approved for residential applications; however, if the area is wet or there is a possiblility of physical damage, Type THHN, THWN or TW conductors run in conduit would be a better choice. Oil-resistant Type SJ (either Type SJO or Type SJOO) has been used successfully on many small installations although it is not specifically code-approved for this use. In either case, the final connection to the generator from its fused disconnect switch should be made with flexible, liquid-tight (plastic-coated) metal conduit. The disconnect should be mounted next to the generator and the conductors should be sized to handle the generator's maximum output current. Output conductors are run from the generator's output terminals to the disconnect switch and then from the disconnect switch to the transfer switch. Output conductors are also run from the inverter and connected to a second set of input terminals on the transfer switch.

The mounting of the system's transfer switch depends upon whether it is manual or fully automatic. If it's a manual unit, it should be mounted in an accessible place where it is easy to operate. If it's an automatic one, the location is determined by proximity to the main service panel. The output of the transfer switch is then wired into the home's main AC service distribution panel. The same conductors can be used from the transfer switch to the service panel as were run to the transfer switch from the generator. In residential applications, Type NM Romex is okay as long as it can be protected from moisture or damage.

The number of conductors required for these wire runs between the generator, transfer switch, and service panel will depend on the type of equipment you have chosen. If your system's AC voltage is 120 volts, the wire run will require an active, a neutral, and a ground conductor. If your system's AC voltage is 240 volts, four conductors will be required: two active, a neutral, and a ground.

AC input to the battery charger should be fed directly from the generator's fused disconnect switch. This is done whether the unit is automatically or manually controlled. Its purpose is to keep the battery charger off-line unless the generator is in operation. Selection of the AC supply conductors to the battery charger is done in the normal manner according to maximum anticipated amperage. The battery charger will require disconnect switches on both the AC and DC sides of the unit.

Fusing and/or a circuit breaker is also recommended on both the AC and DC sides of the battery charger unless the unit is adequately protected internally. Careful attention is required when sizing the conductors that will carry the charger's DC output to the battery bank, because these wires will carry heavy current. This connection must be made on the system side of the battery bank's main disconnect and catastrophic fuse protection. Unlike the hazard posed by the combined operation of the generator and inverter, there is no danger from having the photovoltaic array start to deliver DC power to the battery bank while the charger is operating. Fortunately, DC power is always compatible with DC power as long as it is the same voltage and polarity.

If the generator and transfer switch are automatic, control wiring of the appropriate size will need to be run from the signal output of the system's master controller to the generator's automatic start connections and to the transfer switch. Photron and other PV equipment suppliers have begun to offer integrated control systems that will provide generator start-up and power transfer from one unit. If you have run EMT or RMC from your generator to the transfer switch for your power conductors, the code will also allow the installation of your control wiring in the same conduit.

Wiring the Stand-Alone Inverter

The next phase of SA system wiring depends upon whether the service to loads is AC or DC. If it is AC, power output leads must be run from the batteries to supply DC input to the inverter. The inverter should be located as close as is practical to the battery bank and the wire run should be oversized and as short as possible (using good wiring practices). The wire run from the batteries to the inverter is going to carry some heavy current and the selection of wire size for this run is of critical importance if unwanted voltage drops and overheating of the conductors is to be avoided.

A DC-rated disconnect will be required for the inverter, even though the inverter is likely to have its own built-in power on/off switch and circuit breaker. This is because the NEC requires the ability to isolate a piece of equipment within a circuit in such a way that all power can be disconnected from all current-carrying conductors. The reasoning behind this rule is simple. If the main inverter on/off switch is thrown, there is still live direct current within the unit (ahead of the switch and in the conductor of opposite polarity to that being switched) that could cause injury to a serviceman working on the unit. The circuit breaker is for protection on the load side in the AC line and has no effect on the DC side.

The inverter AC output is fed into a standard AC house distribution panel sized for the amperage required. From there, the wiring is installed in the same manner as in a home that is connected to the utility grid. An exception to this is when there are multiple "dedicated" inverters serving specific loads whose circuits must be kept separate.

DC Distribution

In an all-DC installation, or for the DC service of a mixed AC/DC home, the type of wiring used depends primarily upon the voltage of the system. Since the DC voltage of your system is likely to be lower than the standard AC line voltage, the current required to deliver comparable power will be proportionally greater. In chapter 2, we learned that low voltage and high amperage in a DC system can combine to create a requirement for heavy and expensive wiring. With the exception of fixed lighting and some DC motors (mostly found in well pumps), the majority of DC-powered appliances currently available operate at 12 VDC. If they are to be combined into a moderate-to-heavy overall load—something over 1 kilowatt—amperage draw will increase to the point at which impractically heavy wiring is the only alternative to voltage drops. For this reason an inverter and AC distribution are recommended for PV systems serving a load of more than 1 kilowatt. (Actually, 500 watts might be an even better upper limit for 12-volt systems.)

If you do choose DC distribution, remember that the gauge of wire you use must be sized to safely accommodate the draw of the loads without excessive voltage drops. The same concerns mentioned above for sizing the main DC power feed to the inverter also apply to the main DC power feed to the DC distribution panel. Be certain to carefully label all of your wiring runs and observe the color code with respect to polarity. This is especially critical if both AC and DC distribution circuits are being installed together.

In addition, be certain to use DC-rated switches and circuit breakers for all DC circuits. When a DC circuit is broken under load, an electrical arc is created that can damage switch contacts. When an AC circuit is broken under load the same arc is created. However, it is immediately extinguished within one one-hundred-twentieth of a second as the current wave form swings through the zero point on its path between the positive and negative part of the AC cycle.

The standard wall switches and panel board circuit breakers are designed to handle only AC current. If they are installed in a DC circuit, the best that can happen is they will seem to work normally and will just fail very quickly, whereas the worst that can happen is that your house may burn down. Using DC-rated devices for DC service is vitally important. Unfortunately, they are expensive. Be prepared to pay as much as ten times more for DC-rated devices than for their AC counterparts.

Another important consideration for those who wish to wire their homes for DC power is the selection of plugs and receptacles. The use of standard AC wall receptacles for delivery of DC power is unwise and illegal. Utmost care must be taken to ensure that no AC appliance or equipment is ever connected to your DC power source. With the exception of certain vacuum cleaners and other tools and appliances that use "universal" motors that can operate on either AC or DC, the best you can hope for if you connect your AC equipment to a DC power source is to ruin it. The worst you could expect is electrocution.

The code has not yet established any particular plug and receptacle configuration for use with residential DC power distribution, so the actual plug configuration is up to you. There are many different types of "non-conventional" plug and receptacle configurations available and many are probably stocked at your local electrical supply house. Also, photovoltaic equipment suppliers like Solar Electric of Santa Barbara and Photron have a complete line of DC-rated switches, plugs and receptacles, and DC panel boards, or "load centers," especially selected to be compatible with home DC power.

Maintenance of the Photovoltaic System

The photovoltaic energy system and the components that make it up represent a triumph of solid-state electronics. The absence of moving parts and of mechanically replenished fuel supply or combustion render PV installations virtually maintenance free—or at least relegate most maintenance to the category of preventive care. When storage batteries and auxiliary generators are involved, of course, maintenance is somewhat more demanding, but it is still far from complicated.

Let's begin a review of maintenance procedures with the array itself—surely the simplest part of the system to take care of once it has been properly installed. The most important external factor in the efficient operation of photovoltaic modules is the unimpeded transmission of sunlight. While you can't do much about atmospheric interference from occasional haze or suspended particulate matter, you can get rid of dirt and dust that settle on the module surfaces and cut down on the amount of light reaching the cells.

Usually, normal rainfall will be sufficient to wash settled dust from any array with more than a 15-degree tilt from the horizontal. If rain is infrequent or the array is located in an area of heavy particulate pollution or dust buildup (such as near a heavily used dirt road), you may have to help nature along by doing the cleaning yourself a couple of times each year. Ordinary dust can be gotten rid of with the stream from a garden hose. Heavier accumulations of grime will require washing with soap and water. For rooftop arrays, use a wooden or fiberglass ladder to reach the eave and work with a soft mop or sponge on an extended handle. Then hose down the array from below. Do not use abrasive brushes or cleaning pads or any abrasive cleanser; mild soap will do just fine. If you scratch the glass surface of the module, you will be creating a problem far worse than dirt. The light-transmitting quality of scratched glass is permanently reduced and the cells beneath will never be able to function at top efficiency. Most important of all, make certain that the array disconnect switch is in the "off" position, disconnecting both the positive and negative power leads, before you attempt to clean the array.

Little else remains to be done in the way of array maintenance. If you have a rack-mounted or integrally mounted array, you may wish to occasionally check the security of the electrical connections behind the modules. And regardless of the mounting procedure used, it's a good idea to make a semiannual inspection of the space beneath to make sure that the mounting hardware is tight and there are no leaks at the point where the mounting hardware enters the roof. If you have an integrally mounted system, check for leaks between the modules. Use silicon-based caulking on a dry day to correct minor problems.

Maintenance of power-conditioning equipment—inverters and controls—is not for the layman. The solid-state construction that makes modern inverters relatively trouble free also makes it difficult for anyone other than a trained professional to diagnose problems when they do occur and to make whatever repairs may be necessary. Terrence Paul, in *How to Design an Independent Power System* (Best Energy Systems for Tomorrow, 1981), looks toward the day when all inverters have modular components that can be easily replaced. But that begs the question of diagnosis, which is in any event necessary before the consumer can know what modular part to order and plug in.

By far the best policy with inverters and controls is preventive maintenance, beginning with the first stages of load estimation and sizing of power-conditioning equipment. *Don't overload an inverter.* Regardless of whether you call it underestimating the inverter or overestimating the load, it's a good policy to install an inverter sized to provide about 20 percent more output than you think you'll need.

Beyond this elementary precaution, the best thing you can do for your power-conditioning equipment is to leave it safely alone. Install it in a clean, low-traffic part of the house (preferably in its own closed area or utility room), keep inverter cabinet doors closed and locked, keep air circulation paths free, and make it clear to children and curious strangers that this equipment is not to be tampered with.

Battery Maintenance

Battery storage banks are the only photovoltaic system components requiring a regular schedule of inspection and maintenance. It all comes down to two specific areas: electrolyte replenishment and cleaning.

As batteries are discharged, the water component of the electrolyte solution in their cells is diminished. A typical lead-calcium battery will require the addition of water anywhere from one to two times each year, depending on the electrical capacity of the battery, the volume of electrolyte it holds, and the use patterns it sees. Check your batteries more often, however, especially during their first year of service. Don't overfill when adding water to the cells, and remember to always use *distilled water.*

The schedule for cleaning batteries is not determined by any predictable formula but by simple visual inspection. Clean away the acidic film that may form around the tops of batteries; it's conductive and can sap current, as can other forms of grime and dirt. If you notice a buildup of corrosive white powder around battery terminals, disconnect the batteries and clean the terminals with a solution of baking soda and water. (An old toothbrush comes in handy for this operation.) When the terminals are clean and dry, reconnect the batteries to each other (but not yet to the electrical input source), make sure all terminal contacts are tight, and coat the terminals with grease or petroleum jelly to protect against further corrosion.

Measuring the state of charge in a battery storage system may be done either by carefully monitoring a DC volt meter while the batteries are in their discharge cycle or by taking readings of the specific gravity of each cell with a hydrometer. The volt meter records the output (discharge) voltage during the discharge cycle. The most accurate readings are obtained after the battery has been in the discharge cycle for at least a couple of hours, dissipating any surface charge that may have been present. At 100 percent charge the output voltage will be approximately 2 volts per cell; at 80 percent, 1.94 volts; at 60 percent, 1.89 volts; and at 40 percent, 1.83 volts.

With the more accurate specific gravity method, a hydrometer is inserted into each cell of each battery. The hydrometer measures the specific gravity of the electrolyte within the cells—that is, the proportion of sulfuric acid to water, which changes according to the state of charge. At 100 percent charge, the specific gravity of the electrolyte in a lead-acid battery should be around 1.280 (the specific gravity of water is 1.000). At 80 percent, it is approximately 1.250; at 60 percent, 1.220; and at 40 percent, 1.200. These values are approximate and will vary from battery to battery. You should consult the manufacturer of your batteries for the values for your particular make and model.

Inserting the hydrometer into each separate cell is time-consuming, but the procedure need not be repeated very often if your system is well designed and quality components are incorporated into it. Of course, if the batteries in a storage bank are of the sealed variety with nonremovable caps, specific gravity readings cannot be taken and monitoring of voltage levels should be relied on instead. On some batteries—the Delco 2000 for one—there is a built-in warning device that tells when a certain depth of discharge has been reached. The Delco 2000 displays a small green dot in a conspicuous window when the state of charge is above 60 percent.

In chapter 4, we referred to an occasional need for what is called an equalizing charge for storage batteries used in conjunction with a photovoltaic system. This is designed to create a uniform charge level in the battery bank and requires a more predictable electrical source to do the charging. This is where the backup generator comes in. About once a month (less frequently if the batteries are not discharged deeply on a daily basis), the generator and battery charger should be activated and allowed to give the batteries a "finishing" charge for 3 hours or so at the lower current range necessary to prevent overcharging. (Most battery chargers will automatically taper off the charge as battery capacity is approached.) This operation actually performs a double service. In addition to equalizing the cell charges, it provides an opportunity for the generator to be test-run and kept in good working order on a regular basis.

A further bit of advice concerning battery maintenance has to do with the depth to which the storage bank is regularly discharged. If a battery is represented as being able to stand 1,500 deep charge/discharge cycles, that is no reason to push it to the limit with each cycle. Yes, it can be drained to 20 percent or so of capacity 1,500 times, but if it is drained to only 60 percent, it may easily hold up for an additional thousand cycles and also remain more efficient because the cycle is taking place in the middle range of the battery's capacity, away from the hard-wearing extreme upper and lower ranges. This is why it makes sense to be generous in sizing battery storage and take your compensation in the form of increased efficiency and longer battery life, unless the cost per kilowatt-hour of cycled storage comparison outlined in chapter 4 indicates otherwise. It is common to install an alarm or automatic shutoff device to halt the drain on batteries when a maximum discharge point is reached.

The foregoing is by no means the final word on battery operation and maintenance, both of which are the subjects of two centuries of experiment and experience copious enough to fill several volumes. Nor is it intended to make taking care of batteries in a residential photovoltaic system sound like a full-time job, which it is not. As I have said before, your solar electric system will likely cause you far fewer problems than your automobile.

Few other maintenance procedures remain. If you are using an auxiliary generator, carefully follow the manufacturer's instructions for routine lubrication and servicing, change filters regularly, and keep in mind the anticipated length of the unit's remaining service based upon the length of time it has already been run. Schedule overhaul or replacement of generators in advance of their expected demise (although machinery can always surprise you) so that you will have auxiliary power when you need it.

Now sit back and let everything hum along according to plan, content with the fact that your electricity is being produced from sunlight by your own utility system which, in all likelihood, will be more reliable then the utility system it displaces.

Flexibility, inexhaustibility, cleanliness, dependability, safety, usefulness in otherwise inaccessible areas. The litany of virtues ascribed to photovoltaics goes on and on. And all the while the practical reality of this exciting technology draws nearer to becoming a part of our daily lives. Photovoltaics has come a long way since Vanguard I. Its application to residential needs has come a long way even since its implementation in the Carlisle house. The current novelty of the solar electric home is reflected in the question recently asked of a PV homeowner by a utility company lineman: "What are all those panels that look like blue stained glass on your roof?" A few years from now, that question will not have to be asked.

Credits

Photographs and figures not listed below were supplied by the author.

Photo 1-1. Courtesy Solarex Corporation
Photo 1-2. Reprinted with permission from *Better Homes and Gardens*. Copyright Meredith Corporation, 1985. All rights reserved.
Photo 1-3. Courtesy ARCO Solar, Inc.
Photo 1-4. Courtesy National Aeronautics and Space Administration
Photo 1-5. Courtesy Mobil Solar Energy Corp.
Photo 1-6. Courtesy J. Paul Fowler
Photo 1-7. Courtesy Atlantic Richfield Co.
Photo 1-8. Courtesy ARCO Solar, Inc.
Photo 1-9. Courtesy ARCO Solar, Inc.
Photo 1-10. Courtesy ARCO Solar, Inc.
Photo 1-11. Courtesy Solarex Corp.
Photo 1-12. Courtesy Mobil Solar Energy Corp.
Photo 1-13. Courtesy ARCO Solar, Inc.
Photo 2-1. Courtesy Atlantic Richfield Co.
Photo 2-3. Courtesy Heart Interface Corp.
Photo 2-4. Courtesy Photowatt Corp.
Photo 3-1. Courtesy Mobil Solar Energy Corp.
Photo 4-3. Courtesy Specialty Concepts, Inc.
Photo 4-4. Courtesy Balance of Systems Specialists, Inc.
Photo 4-5. Courtesy Surrette Storage Battery, Inc.
Photo 4-6. Courtesy GNB Batteries, Inc.
Photo 5-1. Courtesy Dynamote Corp.
Photo 5-2. Courtesy Dynamote Corp.
Photo 5-3. Courtesy American Power Conversion Corp.
Photo 5-4. Courtesy Kohler Co.
Photo 5-5. Courtesy American Power Conversion Corp.
Photo 6-1. Courtesy Pulstar
Photo 6-2. Courtesy Hydrotherm Corp.
Photo 6-3. Courtesy Photron, Inc.
Photo 6-4. Courtesy Sun Frost Corp.
Photo 7-1. Courtesy ARCO Solar, Inc.
Photo 7-4. Courtesy Bryan Dahlberg
Photo 7-5. Courtesy Bryan Dahlberg
Photo 7-6. Courtesy Michael Lutch
Photo 7-7. Courtesy Michael Lutch
Photo 8-1. Courtesy Rodale Press Photo Lab
Photo 8-2. Courtesy Jan Bleicken
Photo 8-3. Courtesy Malcolm Greenaway
Photo 8-4. Courtesy Malcolm Greenaway
Photo 8-7. Courtesy Alan Schlorch
Photo 9-1. Courtesy Mobil Solar Energy Corp.
Photo 9-2. Courtesy Bryan Dahlberg
Photo 9-3. Courtesy New England Electric System

Figure 2-10. Courtesy Best Power Technologies, Inc.
Figure 2-11 Courtesy Skyline Engineering, Inc.
Figure 2-12 Courtesy Paul D. Maycock
Figure 3-2. Courtesy New Mexico Solar Energy Institute
Figure 3-4. Courtesy Mobil Solar Energy Corp.
Figure 3-5. Courtesy Mobil Solar Energy Corp.
Figure 3-12. Courtesy Mobil Solar Energy Corp.
Figure 3-13. Courtesy Mobil Solar Energy Corp.
Figure 4-4. Courtesy Delco Remy Division of General Motors Corp.
Figure 4-5. Courtesy Delco Remy Division of General Motors Corp.
Figure 5-1. Courtesy Sandia National Laboratories
Figure 5-2. Courtesy Sandia National Laboratories
Figure 5-3. Courtesy American Power Conversion Corp.
Figure 6-3. Courtesy Best Power Technologies, Inc.
Figure 8-1. Courtesy ARCO Solar, Inc.

Helpful Addresses

Array Structures and Fasteners

AMP, Inc.
P.O. Box 3608
Harrisburg, PA 17105
(717) 561–6208

Kee Industrial Products, Inc.
P.O. Box 207
Buffalo, NY 14225
(716) 685–1250

Leveleg
1873 Garden Avenue
Eugene, OR 97403
(503) 683–8342

Zomeworks
P.O. Box 25805
Albuquerque, NM 87125
(505) 242–5354

DC Appliances and Equipment

Dinh Company, Inc.
P.O. Box 999
Alachua, FL 32615
(904) 462–3464

Kohler Company
444 Highland Drive
Kohler, WI 53044
(414) 457–4441

A. Y. McDonald Manufacturing Company
Energy Products Division
4800 Chavenelle Road
Dubuque, IA 52001
(319) 583–7311

March Manufacturing Company
1819 Pickwick Avenue
Glenview, IL 60025
(312) 729–5300

Parker McCrory Manufacturing Company
3175 Terrace Street
Kansas City, MO 64111
(816) 753–3175

Pulstar
P.O. Box 1959
Orange Park, FL 320671
(904) 264–6453

REC Specialties, Inc.
530 Constitution Avenue
Camarillo, CA 93010
(805) 987–5021

Sun Amp Systems, Inc.
7702 East Gray Road
Scottsdale, AZ 85260
(602) 951–0699

Sun Frost
752 Bayside Road
Arcata, CA 95521
(707) 822–9095

Energy-Efficient Heating Systems

Hydron Company
5580 White Creek Road
P.O. Box 186
Marlette, MI 48453
(517) 635–2359

Hydrotherm, Inc.
Rockland Avenue
Northvale, NJ 07647
(201) 768–5500

Lennox Industries
P.O. Box 809000
Dallas, TX 75380
(214) 380–6000

Temp Master
JNJ Manufacturing Company
P.O. Box 12037
Orlando FL 32859–2037
(305) 841–4131

Inverters and Power Conditioners

American Power Conversion Corporation
89 Cambridge Street
Burlington, MA 01803
(617) 273–1570

Best Power Technologies, Inc.
P.O. Box 280
Necedah, WI 54646
(608) 565–7200

Dynamote Corporation
1200 West Nickerson Street
Seattle, WA 98119
(206) 282–1000

Heart Interface Corporation
811 First Avenue, S
Kent, WA 98032
(206) 859–0640

Photovoltaic Modules

ARCO Solar, Inc.
P.O. Box 2105
Chatsworth, CA 91313
(818) 700–7000

Mobil Solar Energy Corporation
16 Hickory Drive
Waltham, MA 02254–1012
(617) 890–0909

Solarex Corporation
1335 Piccard Drive
Rockville, MD 20850
(301) 948–0202

Storage Batteries

Delco-Remy Division of
General Motors
2401 Columbus Avenue
Anderson, IN 46011
(317) 646–2000

Exide Power Systems
101 Gibralter Road
Horsham, PA 19044
(215) 674–9500

Globe Battery Division of
Johnson Controls
5757 North Green Bay Avenue
Milwaukee, WI 53201
(414) 228–1200

GNB Batteries,
Industrial Battery Division
2010 Cabot Boulevard, W
Langhorne, PA 19047
(215) 752–0555

Surrette Storage Battery, Inc.
Box 3027
Salem, MA 01970
(617) 745–4444

Voltage Regulators

Balance of Systems Specialists, Inc.
7745 East Redfield Road
Scottsdale, AZ 85260
(602) 948–9809

Specialty Concepts
9025 Eton Avenue
Suite D
Canoga Park, CA 91304
(818) 998–5238

The following companies are distributors or dealers for photovoltaic equipment.

Atlantic Solar Power, Inc.
6455 Washington Boulevard
Baltimore, MD 21227
(301) 796–8094

Fowler Solar Electric, Inc.
131 Bashan Hill Road
Worthington, MA 01098
(413) 238–5974

William Lamb Corporation
10615 Chandler Boulevard
North Hollywood, CA 91601
(818) 980–6248

Photocomm
7735 East Redfield Road
Scottsdale, AZ 85260
(602) 948–8003

Photron, Inc.
149 North Main Street
Willits, CA 95490
(707) 459–3211

Skyline Engineering, Inc.
Potato Hill Road
Ely, VT 05044
(802) 333–9305

Solar Design Associates
Still River Village
Harvard, MA 01467–0143
(617) 456–6855

Solar Electric
of Santa Barbara
232 Anacapa Street
Santa Barbara, CA 93101
(805) 963–9667

Sunrise Technologies, Inc.
Chapel Street
P.O. Box 506
Block Island, RI 02807
(401) 466–2122

Bibliography

The publications in the following list that were prepared under contract for government agencies may be ordered from: Superintendent of Documents, Government Printing Office, Washington, D.C. 20402.

Block IV Solar Cell Module Design and Test Specification for Residential Applications. DOE/JPL 1012-78-14. Pasadena, Calif.: Jet Propulsion Laboratory, November 1978.

Boes, E. C., and H. E. Anderson, et al. *Availability of Direct, Total, and Diffuse Solar Radiation to Fixed and Tracking Collectors in the USA.* Albuquerque, N.M.: Sandia National Laboratories, August 1977.

Climatic Atlas of the United States. Washington, D.C.: U.S. Department of Commerce, NOAA, Environmental Data Services, 1979.

Development of Photovoltaic Array and Module Safety Requirements. Melville, N.Y.: Underwriters Laboratories Inc. (Prepared for Jet Propulsion Laboratory, Pasadena, Calif., under JPL contract No. 955392.) June 1982.

Forman, S. E., and D. N. Klein. *Safety Inspection Guidelines for Photovoltaic Residences.* Lexington, Mass.: MIT Lincoln Laboratory. (Prepared for the Solar Energy Research Institute under contract No. SS-1-1327-1.) January 1982.

Gupta, Y. P., and S. K. Young. *Design Handbook for Photovoltaic Power Systems.* SAND 80-7147/1. McLean, Va.: Science Applications, Inc. (Prepared for Sandia National Laboratories, Albuquerque, N.M.), October 1981.

Knapp, Connie L., Thomas L. Stoffel, and Stephen D. Whitaker. *Insolation Data Manual.* SERI/SP-755-789. Golden, Colo.: Solar Energy Research Institute, October 1980.

Kusuda, T., and K. Ishii, *Hourly Solar Radiation Data for Vertical and Horizontal Surfaces on Average Days in the United States and Canada.* Washington, D.C.: U.S. Department of Commerce, NBS, Institute for Applied Technology, April 1977.

Landsman, E. E. "Loss of Line Shutdown for a Line-Commutated Utilty-Interactive Inverter." COO-20279-128. Energy Systems Engineering Technical Note. Lexington, Mass.: MIT Lincoln Laboratory, March 1981.

Lovins, Amory B., and L. Hunter Lovins. *Brittle Power.* Andover, Mass.: Brick House Publishing, 1982.

Maycock, Paul D., and Edward N. Stirewalt. *A Guide to the Photovoltaic Revolution.* Emmaus, Pa.: Rodale Press, 1985.

National Electrical Code 1984. Quincy, Mass.: National Fire Protection Association, 1983.

On the Nature and Distribution of Solar Energy. HCP/T252-01. Washington, D.C.: U.S. Department of Energy, March 1978.

Paul, Terrence D. *How to Design an Independent Power System.* Necedah, Wis.: Best Energy Systems for Tomorrow, Inc., 1981.

Photovoltaics Technical Information Guide. SERI/SP-271-2452 DE85002924. Golden, Colo.: Solar Energy Research Institute, February 1985.

Russell, Miles C. *Residential Photovoltaic System Design Handbook.* Concord, Mass:. MIT Northeast Residential Experiment Station. (Prepared for U.S. Department of Energy under contract No. DE-AC02-76ET20279.) April 1984.

Solar Radiation Energy Resource Atlas of the United States. SERI/SP-642-1037. Golden, Colo.: Solar Energy Research Institute, October 1981.

Solar Radiation Input Data. Madison, Wis.: University of Wisconsin. 1966.

Stobaugh, R., and D. Yergin. *Energy Future.* New York: Random House. 1979.

Stolte, W. J. *Photovoltaic System Grounding and Fault Protection Guidelines.* San Francisco, Calif.: The Bechtel Group, Inc. (Prepared for U.S. Department of Energy under contract No. DE-AC04-76DP00789.) February 1985.

Index

Rodale Press, Inc., publishes RODALE'S PRACTICAL HOMEOWNER™, the home improvement magazine for people who want to create a safe, efficient, and healthy home. For information on how to order your subscription, write to RODALE'S PRACTICAL HOMEOWNER™, Emmaus, PA 18049.